HUMAN DYNAMIC WHOLENESS

(Collected Essays)

by

Sam Razali

authorHOUSE®

AuthorHouse™
1663 Liberty Drive, Suite 200
Bloomington, IN 47403
www.authorhouse.com
Phone: 1-800-839-8640

ISBN: 978-1-4343-7880-4 (sc)

Printed in the United States of America
Bloomington, Indiana
This book is printed on acid-free paper.

Dedication

I dedicate this book to the Great and Suffered People of Iraq who have been through a catastrophic war, and a war that was totally unjust, unnecessary and above all absolutely illegal. One day the historians will write that the invasion of Iraq was the biggest political blunder and most tragically futile war in modern history. Furthermore, it is the only war that was based on deception: *lies, distortions and of course plenty of ridiculous exaggerations.*

Sam Razali

Acknowledgement

The authors of books in writing *Acknowledgement* mainly express their 'heartfelt gratitude' or 'sincere gratefulness' to all the individuals who have helped them, in any way, in writing their books. If I want to do this I need a list of at least 950 names of people who have been of help to me in acquiring the required knowledge for writing these essays. These people consist mostly of my teachers, lecturers, tutors and especially the authors of all the books and articles that I have read over many years. To do so is neither feasible nor commonsensical.

I, however, thank two people, and I thank them wholeheartedly. One of them is Phyllis Higgins, a dear friend, who allowed me to use anytime I wanted the 'special room' (intellectually stimulating and spiritually inspiring room) in her lovely house in Westminster, London. Phyllis, I greatly appreciate that.

The other person is a great and good friend of mine, Frank Peacock, who is a computer scientist. He has never hesitated to be of help to me whenever I needed the kind of assistance that needed technical/computer skills. I have also learned from him a lot of things that I find so useful in my research works. Thank you Frank.

Sam Razali

Preface

I always wanted to write a book consisting of essays or articles on some of the areas that I have researched in. So, let me say in a few words how my research work began.

Many years ago, after finishing a course in psychotherapy, I began doing a thorough research in *Meditation* -------- different systems and various types of meditation; this was my first research area. The main purpose for such study and research was that I wanted to write a book on the relation (if any) between *Psychotherapy and Meditation*. The subject matter in question was not unfamiliar to me, as I already had some knowledge of meditation and I had also practised meditation, more or less, regularly for a while. I soon realized that it was an endless field of study. Though, due to certain circumstances, unable to write the book, I benefited a great deal from the 'adventure' that I had embarked on. The nearly two years of investigation made me more spiritually aware and gave me a better understanding of *mystical spirituality* and a clearer insight into myself. For me it was, what psychotherapist call 'personal growth'. Furthermore, I learned a lot about human being and human mind; in fact it supplemented my studies in psychology. The immenseness of human mind has astonished me and the extraordinary power of the *Mind* has amazed me. The effect of the mind over 'things' is incredible and especially its influence on the body is indisputable. It was this realization of the power and influence of the human mind that led me to the study and research into *Psychosomatics*, which I define as *the science of the interrelation of Mind and Body particularly the influence of mind on body*. I have written a book on the psychosomatic aspects of cancer, which, I hope, will be published in near future. It is said that great majority (in my view at least 95%) of people have not realized the *Greatness of Mind* and have not awakened to the *Dynamic Resources* of this multi-dimensional being called *Human Being*.

Since five or six years two things have been happening to my works: a) my therapeutic work gradually is being taken over by my research work; b) my research work has been expanding from psychological areas into a vast range of fields of study and extending into 'beautifully unknown' horizon.

This book, which contains sixteen articles (essays) on a variety of subjects, caters for people from several academic professions; however, the book is also for non-academics. In fact, most people will enjoy reading many of the topics and discussions covered in this book; they will find them interesting as well as informative. Incidentally, the discussions are not all on psychological topics and health-related areas; some of them are completely different, for example, 'Hegel and his Dialectics', 'The Story of Quantum Mechanics', 'Simone Weil' and more.

CONTENTS

1

Human Dynamic Wholeness:
A Cosmic Concept of Human Being

I am glad that in this and some other essays in this book

I can use and use freely and confidently and at times even lavishly words such as *Consciousness, Self, Soul,* and *Spirit.*

I say this for such words a couple of decades ago (when radical behaviourism and philosophical materialism were still popular among probably most psychologists and scientists) were considered 'taboo'; even the word *Mind* was used with some kind of hesitance or with reluctance. It was a *mindless psychology and a soulless science.* A complete transformation in this regard has taken place. Credit for such transformation, demise of materialism, must be given chiefly to the physicists, the material scientists. The great theoretical physicist, Paul Davis, says,

"It is fitting that physics, the science that gave rise to materialism, should also signal the demise of materialism".

This is my concept of human being; it is a Cosmic Concept. It is what I conceive of Human being as a 'being' in this infinite universe. To put it more ontologically, it is how I view the being of human being in the Cosmos, and especially the relation, *The Link*, between human being and the Cosmos, this boundless unitary system, infinite Universe, of which we are inseparable parts. My concept of human being is mainly the product of my many years of studies and researches in different areas, all related directly or indirectly to human being and human existence in this world. It is, however, my interest and involvement, in recent years, in the field of 'Science and Spirit' that have prompted me to summarize 'the whole thing' in a short essay such as this, *'Human Dynamic Wholeness'* which is also the name of my Website dedicated to 'Science and Spirit'. I believe the 'right concept' of human being and the 'right conception' of human existence in this World is essential to our study of Consciousness and our understanding of the Universe and of course to our progress in bringing together the realms of Science and Spirit.

Human being, as a whole person, is a multidimensional being, who is a self-contained bio-system unified in mind, body, spirit and the subtle energies, and included in these are mental, emotional and vibrational levels. Every one of these is real - it exists; they are not some abstract or imaginary things. And all these levels and dimensions in human being are interrelated and between them (mind and body in particular) there is an ongoing dynamic interaction. Between mind and brain, it is the mind that has the primacy. Mind is capable of exerting extraordinary influence, and usually it has tremendous effect, on body. Brain's influence on mind is limited and mostly conditional. Mind and brain work together, interdependently, as one 'whole'. In order for mind to do its 'job' it needs a functioning brain, and without mind there can not be a functioning brain. This is according to my Mind-Brain Theory (please read it).

As for 'consciousness', I maintain, each human being, individually, does not have his or her own consciousness. We human beings, and also all other living creatures, according to my concept of human being, have in common the 'Cosmic Consciousness'; it 'only seems' to us that every one of us has his or her own (as though born with a) consciousness. In my view, the interrelation, *the interplay*, of the Cosmic Consciousness and a person's brain makes up 'the consciousness' of that person. Human brain, which is the most developed among all the brains of living creatures on earth, is not the 'generator' but a kind of *receiver* (or a sort of 'detector') of Consciousness. As I see it, the difference between two persons' consciousness is the difference between *the interplay* of their brains and the Cosmic Consciousness. And the difference between these interrelations or interplays rests on the neuro-physiological differences of the brains of the persons. This means each person's brain has a different, unique, relation with the cosmic consciousness. That is why I believe (based on this concept) that the consciousness of identical twins is (very much) similar because of the (big) 'similarity' between the functioning of their brains neuro-(chemically, genetically, physiologically etc).

From what has been said in above paragraph we can deduce that human being is 'linked' (literally connected) to the cosmos through Cosmic Consciousness or rather through the interplay of the brain and the Cosmic Consciousness. However, this is not the only *Link* between human being and the infinite universe. Every one of us is also 'connected' to the Cosmos through the *subtle energies*, human bioenergies; this is not the place to discuss human subtle energies. And what about the *Spiritual Connection* with the Infinite? I will say a few words on this in a moment, when I speak of the *Self;* Incidentally, the term 'human consciousness', must not be confused with the 'Self', which is an entirely different thing. Every human being is born with a Self, (or Soul), which now I am going to talk about briefly.

Human being has a 'self-conscious mind' - a mind that is conscious of itself. Human mind is aware of its existence and conscious of its ongoing activities. Animals don't have self-conscious mind. For example, a dog is only aware of its surroundings, whereas I am aware of my surroundings plus the fact that I am aware of my awareness of my surroundings. The cat 'only knows' the meat is on the table, but I know the meat is on the table, and also I 'know that I know' the meat is on the table. In short, animals 'only know' but human beings 'know that they know'. This 'self-consciousness' of mind arises from human being's 'extra dimension' of consciousness (actually it is not consciousness, but a *special dimension*). This 'special dimension' is exclusive to human being. This is our unique possession, the spiritually-related dimension called 'Self' or 'Soul', the centre of our being, from which the ' I ' emerges. It is the 'Self' that gives rise to *Self-conscious 'I'*, thus, rendering human being capable of doing things that animals cannot do: to imagine, to think and to think in abstract terms, to devise and to design, to create and even to create masterpieces. It is only human beings who can 'stretch' their minds beyond their immediate surroundings and their present time into far distant places and into distant past or future even to the infinity. Staunch biologists and over-zealous neuroscientists will say this is due to the higher development of the brain in human being. I say to them, "yes, it may be only in part due to that, but mostly, 'essentially', it is on the grounds that I have just stated". And after all, as has been said earlier, Mind and Brain work inseparably together ----- as a unit, 'one whole'.

But what about the *Mind*? Is human being born with a mind? The answer to this seemingly easy question is 'it depends what we mean by mind'. If by 'mind' we mean all the mental activities, cognition, emotions and emotional feelings, then the answer is "No, we are not born with a mind; we acquire these in this world, from the day we were born till the day we die". The use and usage of the term 'mind' often vary from language to language and they also vary in different disciplines and in some instances among different cultures and communities. Most of the conventional, particularly the ancient, philosophers have used the word "Soul" for the term *mind* usually (perhaps always) with a spiritual connotation.

Human beings probably are the 'unique cosmic beings' in having an amazing power of mind including some astonishing healing resources. We, human beings, are not the victims of our illnesses and diseases, nor are we at the mercy of our genes. Biological pundits, particularly the over-zealous geneticists, in the past three decades or so have been trying hard to convince us that our health and well-being depend on the working of our genes and even to a large extent our lives are determined by our genetic make-up. I say: "Do not believe them". They come up with their 'bankrupt concept' called 'genetic determinism' and continually perpetrate falsehoods. These 'biologically dogmatic' pundits are the 'prisoners' of their materialistic prison. They and many other shallow-thinking scientists find it difficult to swallow the fact that materialism, philosophical materialism, is effectively dead.

Anyhow, let me come to the point. Of course, there are some illnesses and certain diseases that are in part, or perhaps largely, genetic in origin. The truth, however, is that the so-called 'genetic influence' has been exaggerated out of proportion; moreover, with some biologists it has become a 'serious obsession'-'genetic obsession'! It is also noteworthy, even very important, to remember that the genes are not as 'rigid and dogmatic' as most biologists (particularly geneticists) are. The working of the genes to a large extent depends on their environmental conditions. The recent studies and new researches in better understanding of the working of the genes have come up with some interesting facts about genetic influence. I am referring to the science called *Epigenetics*, which deals with (and in the process reveals a lot of 'truths' about) genes and genetic influence.

We human beings (and here I speak mainly, but not only, from my professional experience) are capable of having a great deal of control over our lives; even we can, through the training of our minds, influence and have a considerable degree of control on our health and well-being. In human being there is a great capacity for 'self-healing' through which we can restore health in ourselves. We are equipped with a mind whose power, sadly, still is unknown to the great majority of people. Potentially, we are able, to a large extent, to shape our future; this is not easy, but it is feasible. Human beings must go on discovering their healing resources and powerful capacities, which have to be invigorated and kept fully functional. There is no point in complaining life is tough; of course it is tough. Indeed, life is hard, but this does not mean we cannot do anything about it. And there is no point either in grumbling or mumbling that life has no meaning. It is our job, our duty, to give Life some meaning. A person must 'endeavour' as far as feasible to make his or her life meaningful. All the 'facilities' are given to us for such 'endeavour'. Meanimg reveals Itself to those who seek It and truly love to see It revealed.

To sum up, human being possesses an exceptionally developed brain which is 'in touch' with the Cosmic Consciousness; there is a constant interplay between them. Mind and brain work interdependently as a unit (One Whole); however, between them mind is the 'primary'. Human being is also born with a bioenergy system; this subtle energy is constantly flowing throughout the body, and is "linked" with *Cosmic energy*. Human beings have an incredible 'power of mind' including healing resources; however, these have to be kept active, in good shape, and functional. Furthermore, a person is also born with an 'exclusive dimension' central to which is *Self* from which, as a gradual process, the "I" stems. Self (Soul) is capable of reaching higher levels of Consciousness, even capable of attaining to "The Infinite". Let me say a few words more on this.

We human beings possess a Spiritual Dimension whose centre is Self (Soul), through which we are capable of transcending this phenomenal world and attaining to higher dimensional realms. This is a spiritual journey, which, if followed with dedication and commitment, becomes the 'Mystical Path'. The Path involves uphill tasks and many pitfalls, and at times even trials and tribulations. All these, however, are essential to the preparation of the Soul, *floating in the Cosmic Ocean, to attain her final Goal -- 'Union with Ultimate Reality'. And this is the 'enthronement' of the Soul in the 'mystical eternity'. Such 'Mystical Union' takes place when the Soul becomes face to face with the dazzlingly and here and there glitteringly Shining Light.*

Before the Creation this Light is. It is Alap and Taw, the Beginning and the End; yet, before the Beginning this Light is.

2

Wholeness and Reality

It is essential, in order to talk about wholeness and reality, first to say a few words about the concept of holism (holistic / wholistic concept). Before doing so, however, I should clarify one thing: the meanings and the origins of the two words 'holism' and 'wholism'. 'Holistic' and 'wholistic' are their adjectives respectively.

The words 'holism' and 'wholism' mean, more or less, the same thing: 'unbrokenness', 'completeness', 'unfragmentedness'; their adjectives have the same meaning. Etymologically not only are the two words completely different, but also they are from two different languages. The two words Wholeness and Wholism come from the word *whole*, which is derived from the word 'hal', an old English word; 'holism' comes from the word 'holo', a Greek word, meaning, whole, complete, unfragmented and intact.

Holism, as a philosophic concept, views the universe as 'undivided whole', an unbroken oneness, 'one whole', and in order to know the universe or to have some understanding about its nature, it has to be studied as a 'Whole' and not as separate parts and isolated objects. That is, the reality of the universe has to be and is in its Wholeness. This is because the concept of wholism / holism maintains that the reality of a thing is in the wholeness of that thing, not in its parts or fragments. This philosophical concept is Eastern in origin; later Greek philosophers accepted it. Holistic concept is still shunned by the Western scientists; however, in the past few decades wholism has tantalized the minds of some scientists and many thinkers from different fields of knowledge. Some of them are toying with it and some flirting with it and a number of them accept it though not so courageously. In the West the philosopher who truly had a wholistic mind was Hegel, the German 'philosophic tower'. To Hegel the 'whole', and only the whole, is real, and the attainment to the whole (by moving upwards to higher stages) is central to his well-known *Dialectical Process*.

There is, however, another meaning to the word 'whole'; it means 'healed' or 'holistically healed'. Therefore 'wholeness' also means 'the state of being healed', 'holistic health'. We can see, in an implicit way, a connection between the two meanings of the word in question 'wholeness': a) 'the state of being unbroken', b) 'the state of being holistically healed'. However, it is the first meaning ('unfragmented condition' or 'the state of being unbroken') that I deal with in this essay with relation to 'reality'. But what do we mean by 'reality'?

The word 'reality' refers to the actuality of a thing, and denotes the existent factuality of it. Reality means *the state of being real,* and 'real' means 'actual' and refers to a fact. 'Reality' refers to the state of actuality, factuality, existence, or the 'genuineness of a thing'. Wholeness and reality are inseparable. Is it because genuineness and wholeness go hand in hand together? Well yes, this is only part of it.

In the 20th century there has been a growing interest among many scientists from a wide range of disciplines (notably among Quantum physicists) who have adhered to the philosophy of holism and to the holistic concept --- that 'whole' is real; the parts or fragments of the 'whole thing' are not separate realities of that thing. Put another way, the 'Realness' of a thing is embedded in its *Wholeness*. That is why in order to understand John, as a person, we must meet and associate with him, at least on several occasions, and try to understand him as a 'whole person', for the 'reality' of (about) John is in the 'wholeness' of John: his *psycho-physical entity with all its multi-dimensional interrelations and interactions*. To analyse John's emotional feelings, mental activities, attitudes and his spiritual beliefs separately in order to know him and understand him well, would be futile, misleading and even probably damaging to the psychological status of John. And this also may apply to a wide range of things and aspects in life.

In sharp contrast to holistic concept is 'reductionism'. Reductionist doctrine says, in order to understand the complex things one has to study and analyse them in terms of their isolated parts and simple, basic, constituents. In other words, in reductionism 'the whole' is 'reduced' (hence the word 'reductionism') into parts, basic constituents, which then have to be studied, even analysed, in order to understand its complexity, and to know its 'reality'. It is absurd to claim or even think that reductionism can be applied to every 'thing'. Here, let me say that this essay is not about wholeness versus reductionism, but it is intended to be a discussion on 'wholeness and reality' ---- **that the whole is real, and less than whole is not real; moreover, the parts of the whole are not separate realities of the 'whole'. A 'thing' has one 'reality' only and that is in the wholeness of that thing.**

Perhaps the inseparable relation between wholeness and reality is 'embedded' in the well-known statement which is the 'cornerstone' of holistic concept, and it goes, '*The whole is greater than the total sum of the parts*'. Here, the word 'greater' is the operative word, which has by far a deeper meaning than what it usually means; it does not mean larger or physically bigger, but it refers to some profound aspects of 'the whole'. The 'greater-ness' of *the whole* is due to the interconnections, interdependence, interfusion, and perhaps above all to the dynamic interactions of the parts. The word 'greater' in aforementioned statement refers to greater in meaning, greater in significance and greater in dynamism, and all these collectively give rise to a higher, even probably to the only, reality (the dynamic reality) of the thing in question; and this 'thing' has to be studied as one whole, in its wholeness, in order 'to grasp its reality'. Therefore, the philosophic doctrine of '*holism*' says, as has already been stated that the Universe is 'one whole', a unitary system, and in order to have some understanding of it --- 'to grasp its reality'--- it has to be studied as 'one whole' and not as isolated objects in/of the Universe.

From what has been said, we can, by inference, see that holistic concept is dynamic. In fact, 'the whole', the concept of holism, subsumes the phenomenon of energy. Energy is embedded in the whole (the wholeness of a thing), not in separate parts of the whole. Where there is wholeness there is also energy. We now can see why central to holistic medicine is energy medicine. One of the definitions of *Energy* is 'dynamism' which is related (related in the context of this discussion) to 'wholeness'.

It was Albert Einstein, probably the greatest scientist of all times, who demonstrated what the phenomenon called 'energy' is; before him the world of science did not know what exactly it was. In his well-known theories, among other things, he discusses matter and energy, but it was specifically by his famous equation, $E=mc^2$, that he showed us the nature of *Matter and Energy*. He demonstrated that *Mass* is not made of some sort of 'indestructible substance' that was considered to be the 'basic stuff' of all things. According to Einstein, mass is 'stored energy', and even the two (mass and energy) are convertible. In fact, matter and energy are the two sides of the same reality. Einstein's 'model of reality' (which is vibrant, energetic and ever-dynamic) defied Newton's model of reality, which is known as 'mechanistic'. Our Universe is 'whole', a unified whole----'One Wholeness'. It is the Universe of energies, intrinsically dynamic, one dynamic whole. This wholeness is all interrelations, interconnections and patterns of probabilities---------an integrated Cosmic Web.

Indeed, 'an integrated cosmic web', one whole. The opposite of such wholeness and integration is 'division', 'dismantling', 'disassembling', 'fragmentation', or 'taking apart'. The act of 'taking apart', disassembling, has become a preoccupation with the Western societies, thus causing concern among many scientists from different disciplines and fields of knowledge. It is said that such 'disassembling of the whole' causes confusion, and also some scientists maintain that the perceiving of the fragments of a 'thing' as separate realities of that 'thing' is nothing but an illusion. Perhaps the scientist who held most strongly such views was **David Bohm**, one of the greatest physicists of the 20th century.

David Bohm (sadly no longer with us) was professor of theoretical physics at Birkbeck College, London University; he also was an eminent Quantum theorist. Bohm became notably known for his arguably the most advanced theory of the universe; the theory is called 'implicate order'. He was a friend and in the early days also a colleague of Einstein who liked Bohm and also admired him for his intellect; it is said that he was the favourite physicist of Einstein. He, Bohm, was also well acquainted with Eastern philosophies, and to some extent they influenced him. The late Krishnamurti, internationally well-known Indian spiritual teacher, was a close friend of his. They would frequently meet and discuss some philosophical subjects and spiritual matters.

In his well-known book, *Wholeness and Implicate Order*, Bohm deals, among other things, with 'wholeness and fragmentation', an important topic of his book; he critically discusses it and skilfully argues that the 'reality' is in the 'undivided whole'. What concerned him, perhaps most, about the West and Western way of thinking was 'fragmentation', dividing the whole into parts, in order to find or understand the reality of the things. Bohm saw the main difference between Eastern and Western philosophies to be *the question of perceiving the reality*. In the East the 'reality' is seen in the 'whole', but in the West the 'reality' is looked for in the separate parts and the bits and pieces of the whole. In short, Western scientists love to 'chop' the whole (the wholeness of a thing) into parts and in fragments in order to find out the reality or the true nature of that thing; moreover, they see the fragments as separate realities of the 'thing' in question, something that really annoyed and even saddened Bohm.

Bohm's 'anti-fragmentationism' is not quite the same as 'anti-reductionism'. Fragmentationism subsumes reductionism, but then it goes much further---deeper and wider; it has psychological, societal and philosophical implications. In his book, Bohm forcefully argues against fragmentation and Western 'fragmentary mentality', which, he believes, has confused the Western mind. He says,

> **"For fragmentation is now very widespread not only throughout society but also in each individual; and this is leading to a kind of general confusion of the Mind, which creates an endless series of problems and interferes with our clarity of perception so seriously as to prevent us from being able to solve most of them."**

Indeed, as we can see, 'fragmentationism' did concern Bohm. He really thought of it as a serious problem. 'Fragmentary mentality', as Bohm puts it, "...is leading to endless series of problems and interferes with our clarity of perception so seriously..." and also such problems have their adverse effect on political, religious, and the societal issues. He says,

"Then, society as a whole has developed in such a way that it is broken up in separate nations and different religious, political, racial groups etc. Man's mutual environment has correspondingly been seen as an aggregate of separately existent parts, to be exploited by different groups of people."

The notion of wholeness and reality is closely related to Bohm's scientific and philosophic theory of *The Implicate Order*. This is a highly advanced theory of Universe, in which Bohm in order to explain his theory uses terms and special phrases that are both interesting and intriguing: 'enfolded phenomena', 'unfolded process', 'the implicate order of the universe', 'the explicate order of ....' , and so on. He also created concepts such as 'holomovement' and there is also 'holonomy', which are based on or closely related to, the principle of hologram. By the term 'implicate' Bohm means that which is enfolded and invisible or hidden. And by the term 'explicate' he refers to being unfolded, visible and so on. Furthermore, Bohm in his book refers to the Universe as *Holomovement,* a term which denotes 'wholeness in movement', a hologram in action and so on.

Some say Bohm derived his concept of hologram from the holographic model of the well-known Stanford neuroscientist, Karl Pribram. This is not true. However, the two men knew each other well and for a while they collaborated together on their theories especially holographic concepts. Therefore they learned from each other's findings, and also probably in some ways they inspired each other. For example, Pribram was very much impressed when he noticed the similarities between Bohm's Implicate Order and certain ideas that he, Pribram, had of Hologram. Furthermore, Pribram's *Holonomic Model* of the brain, which roughly denotes brain's cognitive model, was developed in collaboration with Bohm's theory. Here it is noteworthy to mention that what drew Pribram to holography was his investigation into memory and the storing of memories in the brain. After his extensive research in this area Pribram stated conclusively that in brain there is no special place where memories are stored, Human memory, he says, is dispersed all over the brain on the principles of holography ----- a complete holographic model. However, it was not Pribram who discovered this. It was Karl Lashley, a distinguished brain scientist who in 1920's found that no matter which part or portion of the brain is removed a memory cannot be eradicated. But Lashley was unable to explain why this is so. A few decades later the other Karl, Karl Pribram, came along and explained that this is due to holographic working of the brain. Pribram was the first to discover the holographic model of the brain ---- that brain functions on holographic principles. He also stated that memories are not stored in the cells, neurons, but rather in the form of wave patterns over the nerves..

The contributions of Bohm and Pribram, considering their original concepts of holography, particularly when combined with Bohm's implicate order, are immense to science and philosophy. In fact Bohm's theory of the Universe, *Wholeness and Implicate Order*, is known and accepted as (arguably) the most advanced theory of the Universe and of course also Consciousness, as he himself said that his second passion, after universe, had always been consciousness. But then every physicist who is the researcher of consciousness has a close affinity with knowing about the universe, and a theorist of the universe is also drawn to the research into consciousness; the studies of the two seem to be inseparable. Bohm's theory tells us the universe that we see and the world that we perceive is not all that there is. He maintains that underlying our perception of this world is a deeper layer, by far of a more profound *Reality*, to which he refers as *Implicate Order*. The universe, he says, operates on the principles of holography. And when he spoke of human brain in relation to the universe he would say "a holographic brain in a holographic Universe". His Universe is vacuumless. To Bohm Space is not emptiness or vacuum; the infinite space in the infinite universe is filled with Plenum, the ground for existence.

In Bohm's classic book, *Wholeness and Implicate Order*, probably the most frequently used words are "Wholeness" and 'whole', and the word "Reality" meant everything for him; in fact he lived all his life as a "Seeker of Reality". David Bohm, the Scientist-Philosopher, was a Realist who lived with a Whole Mind and Wholistic Way of Thinking.

From what has been said in this essay we can confidently say that it is 'good ' to have a wHolistic Mind, and it is 'right' to aim at acquiring a wholistic way of thinking. Persons with such way of thinking and perception see and perceive 'things' differently (more real). For example, they see a Society (any society) as 'one whole' and perceive the *reality* of the society not in its separate groups but in the totality of its groups of individuals and communities as *one whole;* that is, not in its white people, black people, rich people, poor people, Christians, intellectuals, scums, Buddhists, perverts, Moslems, saints, Jews, vegetarians, insomniacs, artists, gamblers etc, but in the combination, amalgamation, of all these groups living and functioning together as *one whole.* This means *society,* as *one whole,* is 'the reality', and the only reality, of that society.

Implicit in Holistic concept is 'something' of a spiritual nature. In fact, underlying the term 'One Whole' is a mystical element. In my view, the mystical mind (also, among other things) subsumes the holistic mind. Furthermore, central to Hegel's Dialectics is the notion that *the reality is in the whole.* I see, from my study of Hegel (this is my subjective view), the dialectical mind is somewhere between mystical and holistic minds. Incidentally, Hegel's philosophy in general, and his idealist dialectics in particular, are 'mystical', though in an implicit way.

3

Psychosomatics
And
Psychosomatic Disorders

Psychosomatics is the science of the interrelation of the mind and body with a particular emphasis on the influence of mind on body. A psychosomatic researcher studies the psychological processes and physiological functions. Psychological processes (which refer to mental activities, emotional states, feelings and so on) are also closely related to and have considerable effect on the human energy system.

Psychosomatic disorders are brought about by the undesirable, unhealthy, psychosomatic processes. The term 'psychosomatic process' refers to the interplay of mind and body particularly their interactions. Psychosomatic interaction does not take place in a direct way but through the bioenergies.

Psychosomatic researchers in the West talk or write about psychosomatics and psychosomatic disorders only in terms of mind and body; they hardly (in fact never) mention human bioenergies. This is like claiming (or pretending) that all that is needed for driving a car is the driver and the car; petrol is not important! According to my psychosomatic theory, the human energy system, the bioenergy in human body, acts, among other things, as an 'interface' between the psyche (mind) and soma (body), or to put it more specifically, between psychological processes and physiological functions.

The concept of the interrelation of mind and body goes back to ancient times. Most belief systems and many philosophies have maintained that soul (mind) has a considerable influence on body. In the East mind and body have always been regarded as two distinct but never as two separate elements. In fact throughout the centuries not only in the East but also in the West there have been philosophers who have talked about or commented on the 'oneness of mind and body'. Let me quote Plato on this; he has said it in Charmides, quoting Socrates:

> **"As it is not proper to try to cure**
> **the eyes without the head, nor the head**
> **without the body, so neither is it proper**
> **to cure body without the soul, and**
> **that this is the reason so many**
> **diseases escape Greek physicians,**
> **because they are ignorant of the Whole."**

The above statement is very much a psychosomatic (mind-body related) statement. It is not so simple or easy to give a single definition of the term 'psychosomatic illness (disorder)' that can apply to all disorders of psychosomatic nature. However, I will attempt and do my best. A psychosomatic disorder is a physical illness or disease in whose development psychological factors and certain psychosocial elements mostly of emotional nature have played (a significant) part. It is estimated, mostly by psychosomatic researchers, that today the great majority of illnesses and diseases (nearly 80%) are, to varying degrees, of psychosomatic nature. Furthermore, as one who has done a lot of study of the psychology of cancer and has carried out research into psychosomatic aspects of this disease, I want to say that great majority of the types of cancer are, at least in part, psychosomatic. Let me be more specific: in the development of nearly all cancers with solid tumour psychological and psychosocial elements play a greater part than all physical elements put together. In such contexts the 'elements' concerned mainly refer to the *stresses and strains* that we experience in our lives.

The term 'Stress' in ordinary language, is the 'pressure' of psychological and psychosocial nature in an individual's life that he or she finds it, and feels it, to be beyond their ability *to cope with*; they find it stressful (in some cases unbearably so) when they cannot *manage* the stresses and strains and the difficulties that they encounter in their lives. Such 'ability' (to cope with or to manage) varies from individual to individual. It depends a lot on how the person concerned perceives the stress in his or her life and also on the 'means or strategies' that they have at their disposal for coping with or managing their stresses and the difficulties in their lives.

In the technologically advanced societies such as Western Europe and America, and of course Japan too, many of the over-ambitious individuals regardless of their personal limitations, social restrictions and all other obstacles in life go on struggling to attain even the unattainable. This kind of ruthless competition and rat-race success is highly detrimental to the health of the individuals and to the well being of the community as a whole; it is a societal malaise. Not an irrelevant example is the fact that until about 30 years ago cardiovascular diseases were rare among people under the age of 45.

Today this is no longer true. For example, coronary heart diseases are common among people in their 40's and even not so rare among men in their 30s. And what about mental illnesses, nervous diseases, drug addiction and the increasing number of suicides, which are mostly stress-related or psychological problems. Our conventional medicine cannot cope, nor can do much about these serious problems. Such 'helplessness' is due to the fact that the Western orthodox medicine is a 'physical medicine'; it is not a healing medicine. It has advanced only technologically. It can, for example, perform most of the surgical operations, and carry out the transplantations of some of the body organs with reasonable success. Yet, it is helpless in curing, or even ameliorating, a wide range of illnesses let alone in doing something positive about some of the diseases that tormentingly decapacitate and finally kill their victims. In short, Western Orthodox Medicine is reasonably successful in dealing with purely physical (physically-caused) illnesses, and helpless in doing something positive about complex or multi-causal diseases; the best example would be cancer. Now we can see why after a century of research into cancer and spending (wasting) billions of dollars on its research we are, more or less, in the same position that we were when we started. Why? Because cancer is complex and especially in majority of types and cases is, in varying degrees, of psychosomatic nature.

What is needed is a 'Healing Medicine' for the 'whole person'. That is, what we need is a 'Holistic Medicine' central to which is energy medicine. This kind of medicine is multi-disciplinary which, of course, also includes the Western technologically advanced medicine. Or, alternatively, we must have 'psychosomatic medicine' made available in all hospitals. Psychosomatic medicine is concerned with the study and research into psychosomatic disorders and with the application of psychosomatic treatment, which is the integration of medical and psychological treatments.

Western traditional medicine has for far too long been based on the Newtonian model of reality, a completely mechanistic concept. it holds the view that the universe is an infinitely huge space containing isolated objects. This was the culmination of the Cartesian mind-body concept (Rene Descartes' concept of mind and body). Such model or concept is the basis of the Western conventional medicine whose view of the patient is that he or she is a physical body containing organs and bits and pieces ---a physical being only. And the patient's illness or disease is seen as a physical condition only (a physically-caused problem). This erroneous 'model' of physical or mechanistic nature ought to be replaced with an energetic, vibrational, model of reality which is the way Einstein viewed the matter and in fact the whole universe. He said that matter is a form of energy, and even mass and energy are convertible. It is only by **a)** following such an energetic, 'Einsteinian', model of reality, and **b)** by accepting the holistic concept of human being that the conventional medicine can be transformed. It is only then that it will become a *'healing medicine' for the whole person who is a multi-dimensional being.*.

Yes indeed, human being is not a 'walking body' or a being of physical dimension only. Human being is a person, a whole person, unified in mind, body and spirit and all the emotional feelings plus the subtle energies to which the field of medicine in the West has always turned a blind eye. These elements (dimensions and levels) are in a state of constant interplay; hence, their interactions, particularly psychosomatic (mind-body) interactions, play a crucial role in the state of our health and well-being. *It is therefore morally wrong and intellectually inconceivable to treat a human being, when sick, suffering from a serious illness (especially a complex or multi-causal disease), like a car that is not functioning properly-----focusing on the 'body repair' only.*

4

Self-Awareness:
Something to aim at

When I decided to spend some of my therapeutic time working with cancer patients, giving them psychological treatment, I had already finished my study and research, as part of a project, into different systems and various types of meditation. Apart from investigating into the spiritual dimension and mystical aspects of meditation, I also studied a large number of researches carried out into the psychological and especially the physiological changes that the regular practice of meditation brings about in the meditators. The chief object of such researches as far as the biological aspects of meditation is concerned has always been to find out to what extent the regular practice of meditation could be considered an 'alternative' to the medical treatment of some of the bodily dysfunctions, illnesses, diseases, or any physiological mal-functions or disorders.

The first thing that the researchers into meditation and related areas usually find out is the fact that the regular practice of meditation renders the meditator able to attain 'self-regulation' or as some refer to it as 'self-control'. Here the term self-regulation refers to the ability of the person to regulate some of his or her psychological processes, and especially certain physiological functions. This means that the person knows and also feels that he or she is gaining a control (a 'sense of control') over some of his or her psycho-physiological processes. Incidentally, here the two terms 'self-control' and 'self-regulation' may be used interchangeably; however, strictly speaking there is a difference between them. The term 'self-control' has a wider implication, mostly in psychological and social (and perhaps in more) contexts. The term 'self-regulation', however, is more technical and in the kind of discussion that I am having here it is more specific; it is associated with regulating (having 'voluntary control' over) certain autonomic processes of physiological or biological nature.

Now let us ask ourselves, "What are the factors or elements in the regular practice of meditation that result in the kind of 'self-control' that I just described. Put another way, is there an underlying 'element' in the systematic practice of meditation that gives rise in the meditator to 'self-regulation', which enables the person (at least to some degreee) *to have access* to his or her autonomic functions or processes? ("having access to" conveys a more accurate meaning than 'having control over' in this context.)

Though in the East since ancient times *voluntary control over the autonomic nervous system* (which is what 'self-regulation' is about) has been considered feasible and also among different groups of people has always been in practice, in the West until 40 or 50 years ago this was unthinkable. Self-regulation comes about progressively as a result of the regular practice of meditation or any meditative technique or self-regulatory strategy. The process, self-regulation, involves two interrelated phenomena, one on a conscious level and the other on a deeper level in the course of the practice over a period of time. But I wanted something simpler (a more straightforward explanation) than this. Was there, I wanted to know, something as, so to speak, 'the corner-stone of meditation'? I wanted to find out if there was 'something' that was fundamentally common to 'meditative techniques' the regular practices of which give rise to the 'fruits' (all the benefits) of meditation ------ psychological, physiological, biological, spiritual and so on.

To cut the story short, I found out, mainly through my meditational and from my professional experiences that the 'thing' I was looking for was what I call 'Self-awareness'. Yes indeed, self-awareness is the main element in all meditative techniques and self-regulation strategies; it is essential to self-regulation or self-control. *Self-awareness is the 'corner-stone' of meditation.* I became more convinced of this 'meditational fact' as I found out that 'the thing' all meditations (irrespective of their origin, nature or procedural practice) have in common is *Self-awareness*. It is the essential element, in fact the 'basis', for leading the meditator to higher levels of consciousness. But what is this 'thing' that I call 'Self-awareness'? After all 'the term' has been used by different people or therapeutic groups with different denotations and with various connotations. Then let me try to define it, or at least to say what I mean by it.

Here, in this meditational discussion, by 'self-awareness' I mean, ***being aware of mind and of body and particularly of their interrelation.*** It is being aware of some of the mental and of certain physical goings-on and especially of their interconnections. A technical, a more specific, definition of the term 'Self-awareness' will be, being conscious of some of the psychological processes and of physiological functions and in particular being aware of the interrelation of such processes and functions. In short, to put it succinctly, *Self-awareness* *is a Person's Psycho-Physiological Awareness.* This kind of 'awareness' (of mind and body) begins in the early stages of meditation, thus, enabling the meditator progressively to acquire an increasing awareness of the interrelation of mind and body, both working as one 'Unit'. ***This is the 'Awareness of the Self'*** ---- *'Self-Awareness'.*

Now we can see the crucial role of Self-awareness in meditation and its capacity in healing or therapy and even beyond. Once I realized this I began to design some methods and special techniques the regular practice of which would enable the person to acquire self-awareness more effectively and more quickly (even than meditation). Initially I devised them especially for cancer patients, as part of my psychological treatment for them. As I progressed in my therapeutic work I developed these therapeutic methods further; this I did mainly by combining them with breathing disciplines and therapeutic imagery. Having realized their therapeutic effectiveness, I then began to teach them to other people who needed therapy. Since then I have taught these 'healing methods', Self-awareness Therapeutic Techniques, to a number of healthy (non-patient) individuals. As I said earlier, Self-awareness is the 'corner stone' of meditation and meditative techniques and the basis for all the benefits thereof. Therefore, I suggest that it should (perhaps 'must') be made the corner-stone of many therapeutic methods; it should be part of therapies including (and especially) part of psychological treatment for psychosomatic disorders.

Let me say at this point that there are in this world some people who have a reasonable degree of self-awareness without having practised meditation or any kind of meditative techniques or self-regulatory strategy. The reason for this is that these people live (or probably have lived at least most of their lives) a 'meditational life' or have had a contemplative way of living, and also their philosophy of life and their attitude to this world and to their fellow human beings are not ordinary, but somehow spiritual.

We can see now why the late Rollo May, the leading humanistic psychologist said, 'Becoming a person means a heightened awareness [awareness of the self]'. This 'heightened awareness' may also be thought of as 'heightened consciousness'. The well-known Danish existentialist, Soren Kierkegaard says, "The more consciousness the more Self". To these sayings, from these two great men, I want humbly to add: *The higher is the Self-awareness, the higher becomes the Sense of Being Alive. Furthermore, central to Self-development is the developed ('heightened') Awareness of the Self.*

Yes, life is hard; to deny this, is to deny the fundamental truth about life; and those who tell us 'life is easy and rosy' are either escapists (can not face the realities of life) or shallow-thinking optimists. It is, therefore, our duty to strive to get to higher stages of Self-development and to attain, a good degree of, what psychotherapists call, 'personal growth', which includes also Self-awareness. I say this because it is only then that we may gradually know what life is about and probably also we can find some meaning to our lives, for our being in this world. Then we will see that in this universe we are not bewildered sheep for slaughter, but human beings with dignity, who are capable of having considerable control over their lives. I believe, I know and I am convinced that we are not the victims of our illnesses. We have the healing resources and an *incredible self-healing capacity* for taking 'good care' of our health problems and even we have deep inside us all the 'required ingredients' for surmounting many, probably most, of our personal problems and difficulties in life.

Indeed, we human beings have by far greater resources and potentialities than we can ever imagine. And by utilizing these *inner resources* we can free ourselves from the shackles of life. The attainment to such States or Stages is not easy, in fact in most cases is extremely difficult; yet through perseverance and commitment it is feasible. Yes, life is hard. If we want to live a life that is somewhat or somehow meaningful we must work on ourselves. Let us then start with aiming at "A Heightened Awareness of the Self", Self-Awareness.

Therefore again I say, let us aim at attaining a high level of Self-Awareness.

5

Hegel and his Dialectics:
The Dialectical Minds
(including Karl Marx)

I want to shout and say to the whole world that Hegel is the greatest philosopher of all, but I am frightened to do so; Platonists or Aristotelians might kill me for saying that! Hegel was a German philosopher who was born in 1770 in Stuttgart, Germany and died in 1831 in Berlin.

Hegel's philosophy is not easy to understand. Hegel, according to most students of philosophy and many philosophers including Bertrand Russell, is hard to understand, and his philosophical works are probably the hardest to follow. It is chiefly because of this philosophical abstruseness and Hegelian abstract thinking that Hegel is not read as widely as he should be; and it is a pity, because his works are philosophically stimulating and intellectually enriching. Furthermore, Hegelian philosophical system, here and there, is 'seasoned' with 'mystical spices', and perhaps this is more manifest in his dialectics and in his dialectical thinking. In fact there are many, perhaps most, Hegelians and writers on Hegel who would say the whole philosophy of Hegel is mystical. I will talk of this later.

Hegel's influence especially in the West has been tremendous; not many philosophers have been so influential in different fields as Hegel has been. He is the idealist to whom Absolute Spirit/Mind is the highest, and also it is central to his philosophy. To Hegel philosophy must be about uncovering the truth, and concepts are to unfold the 'hidden reality'. And the 'reality' to Hegel is only in the 'whole', and nothing less than *whole* is real; this is central to his dialectics.

Hegel's dialectics is the heart of his logic, which is entirely different from Aristotle's logic. It has nothing to do with syllogism; it needs neither 'premise' nor is it concerned with 'conclusion' as such. To Hegel logic is a process of attaining to the truth, or it is a 'tool' for discovering (or uncovering) the *Reality*. This is achieved, in Hegel's dialectics, through a series of contradictions. In Hegel's dialectics such clashes and contradictions give rise to the exposition of conceptual unfoldings.

Those who know something about Hegelian dialectics most probably also know about the three oft-repeated terms: Thesis, Anti-thesis, and Synthesis. Hegel never used these terms; they are the creation of Hegelian translators who invented these terms, most probably, in order to facilitate the process for the readers of Hegel to understand his idealist dialectics --- and in a way they did a good job too; it works.

Hegel is not the inventor of dialectics but the discoverer of it. The dialectical process has been in existence since the beginning of Existence. Anywhere there is action, movement or something going on, or whenever there are interactions of minds, the dialectical process is at work. In fact, the world and even the whole universe operate in a dialectical way. And the ultimate purpose of this 'cosmic dialectics' is *the attainment to Absolute Idea(lism)*, which is all that matters in Hegelianism; anything outside the reality of the 'Absolute Idea' is unreal, painful, transitory, insubstantial --- a delusory world.

Hegel's dialectics, dialectical method, is a process of the exposition, the unfolding, of 'reality'. It is an upward movement to stages of higher order, and each stage, through such movement, gets closer to the 'whole' which, according to Hegel, is 'the only reality'. Let me explain it briefly and simply and in my own way; I hope this will enable the reader to have a general idea of what Hegel's dialectics is about. To have a general knowledge of Hegel's Dialectics is not so difficult; however, to understand it in depth and to grasp its philosophical profundity, is hard and requires an intellectual struggle over a period of time.

Let us begin with a situation, position, condition or any 'starting point', usually referred to as ' Thesis'. This thesis on its own is not adequate to do what it wants to do; it is probably weak or not self-sufficient. Sooner or later an 'opposition' will emerge to confront the thesis. This opposing element, or counter-force, usually is referred to as 'Anti-thesis'. The thesis confronts the anti-thesis, and both go through their contradictory phases. A problem arises! The problem is that the anti-thesis, too, is not strong enough, not self-sufficient, to be capable of becoming satisfactorily functional. After having been for a while (or perhaps for too long) through their clashes they give up (what else can they do!). They put aside their conflicts and disentangle themselves from their contradictions and silly confrontations; they come together and try to reconcile ------- and reconcile they do. Such *'coming-togetherness'* and reconciliation give rise to their 'unification'. This 'reconciliatory unification' of thesis and anti-thesis is called *Synthesis,* which is a state or a stage of 'higher order', i.e. higher than from where it all started ('the starting point'). But this is not the end. In great majority of dialectical instances, the process of contradictions and reconciliations will go on, and very often on and on. This means that each time a new synthesis is obtained, it (the synthesis) becomes a 'new thesis', a new starting point. So the process will go on being dialectically repeated; thus, the stages of ascending, higher, order will continue to be created until the *Whole* has been attained to. This means through the dialectical processes a series of unfoldings take place until the 'reality' (which, according to Hegel, is only in the 'WHOLE') has been uncovered ----- uncovering the reality. Let us remember central to Hegel's Logic is the endeavour to acquire the truth or to uncover the reality.

Incidentally, in Hegelian dialectics the contradictions, negations, and the negations of negations form the 'dialectical interrelations', and these are carried through and preserved, thus remaining as part of the process.

The dialectical process like mystical process is an 'upward movement' (an 'ascending process') to stages of higher level, or higher order. There is somewhat a parallel between them. In Mysticism the practising mystical aspirant (through carrying out the essential practices and the required disciplines, which involve also certain psychological 'ups and downs' and ambivalent experiences) aims at moving progressively to *Higher Levels of Consciousness within the realm of Spirit for the sole purpose of attaining to the State of Union with "Ultimate Reality".* In Hegelian dialectics the whole process aims at moving (through a series of contradictions and opposing elements) to the stages of *Higher Order* for the sole purpose of attaining to "The Whole"----i.e. aiming at *uncovering the Reality,* because according

31

to Hegel 'Whole' is the only Reality; and nothing less than 'whole' is real in its true sense.

Hegel's dialectics sometimes has been called (and regarded by many as) **'the algebra of revolution'**. I believe the first person who coined it thus, was Alexander Herzen. Indeed, Hegelianism in general and the philosophy of Hegel's dialectical process in particular, is somehow of 'revolutionary' nature. Put another way, Hegel's dialectics is, so to speak, a 'road' to *Complete Transformation.* Perhaps it is this 'implicit revolutionary' aspect in Hegel's philosophy that has fascinated many of the 'Radical Minds'. A radical mind that has some understanding of Hegelian philosophy, and particularly a reasonable degree of insight into and also an adherence to Hegel's Dialectics, I call a **Dialectical Mind.** A truly dialectical mind also is necessarily a '(W)holistic mind'. I will come to this later.

Some of the greatest minds of the 19th and 20th centuries were influenced by Hegel's philosophy and were fascinated by Hegel's dialectical thinking. The nature of such influence and fascination has not been always explicit and straightforward. Sometimes, in some cases, it has been unclear and strange --- intellectually complex and emotionally ambivalent. Such complexity and ambivalence towards the Hegelian world is not difficult to understand, for Hegel's world is highly complex and also ambivalent. The world of Hegel and Hegelianism, in general, is not easy to grasp: it is easy to love and easy to loathe, easy to ignore and easy to embrace. No wonder after his death his world was divided into two parts, *the Left (Young) Hegelians and the Right Hegelians.* The young **Karl Marx** was among the Young (Left) Hegelians. Here I am going to talk mainly about a few 'dialectical minds'. The first and foremost has to be Karl Marx, who had a most radically dialectical mind.

Karl Marx's psychological affinity with Hegel and his intellectual relation to Hegelian philosophic system is not a simple one. However, Hegel was the greatest influence, 'formative influence', on the young man called Karl Marx. Furthermore, it would be true to say that he, Karl Marx, lived practically as a Hegelian though he did not adhere to Hegel's Idealism. What I have just said can be deduced (explicitly or implicitly) from Marx's ideas, language and thinking expressed or stated in his writings and from his way of writing. The early writings and language of Marx in both style and syntax are very much Hegelian. Furthermore, Many of the well known 'Marxist jargons' - political, social and economic - are Hegel's. But later Marx developed his own style of writing, and his own language, thus always trying to be original. Though not a Hegelian as such deep in his heart Karl Marx remained a man with great affinity for Hegel. Marx knew well that when it came to his 'intellectual development' he owed a lot to 'his philosopher', Hegel. Am I trying to discredit or belittle Karl Marx? Absolutely not. I am not a Marxist but an admirer of Karl Marx. He was a great man and a genuine revolutionary; he will remain as one of the great men in human history. In some of his major writings we see the originality of his thoughts and the forcefulness of his arguments that he puts forward. Marx is one of the most (perhaps the most) misrepresented men in human history. His name was used and his philosophy is still misused by many of the so-called Marxists and also by a few (especially in the past fifty or sixty years) unscrupulous States in certain parts of the world largely for their 'ulterior motives' or for attaining their corrupt aims and inhumane ambitions. In this respect there is a parallel between Marxism and Darwinism. Darwin's name also has throughout been misused and Darwinism abused by many of the atheistically motivated biologists, and intellectually dishonest geneticists mainly for the sake of achieving their goals.

It is said that Karl Marx stood up Hegel on his head; or he turned Hegel upside down. By this they mean Marx turned Hegel's dialectical idealism into dialectical materialism ---- in short, he changed the dialectics of mind into the dialectics of matter. This is not true - not true at all. Throughout his life, Marx neither in his writings nor in his speeches/lectures used the term 'dialectical materialism'. Perhaps there would be 'some truth' in saying Engels, a close friend and the collaborator of Marx turned Hegel upside down. I say this because Engels was a staunch materialist and all the materialist aspects of Marxism were written by him; however Engels claimed that he read the manuscript (what he had written) to Marx and Marx approved of it ---- be it as it may. But let me also be fair and say that Engels did not use the term 'dialectical materialism' either; in his writings only twice he uses the term 'materialist dialectic' as an opposing reference (counterpart) to Hegel's 'Idealist dialectic'. Here I must say that it was Ludwig Feurbach (a whole-hearted materialist philosopher) who, when it comes to materialist philosophy, had the most influence on Karl Marx. Such influence and its effect, however, gradually diminished as Karl gracefully matured into maturity (I am aware of my tautology).

The truth is that the overzealous materialist Marxists 'shrouded' the name Marx and the term Marxism in the discredited label, 'dialectical materialism', and by their vigorous propaganda they tried to sell the 'package' to the whole world. It was, however, Plekhanov who 'invented' and also largely popularised the term 'dialectical materialism' in Soviet Union where it became the major part of the philosophy of Communism. Such philosophy has nothing to do with 'Socialism', which is a civilized and humane economic system. Socialism ('ethical socialism' based mainly on moral grounds and not purely on ideological ground) aims at running an organized economic system that serves the great majority of the people, even the whole society; it is based on Social Need, not on individuals' greed. Such system, a special mode of production, is administered by relying chiefly on human cooperation and not on ruthless competition (as is the case with exploitative system of capitalism). Socialism maintains that the Society (say, the great majority of the people) are entitled to benefit from the wealth and the economic prosperity of the Socialist Economic System, since the distribution of the wealth is the major object-principle of this kind of organized system of economy. Socialism holds the view that the means of production, ought to be owned by the people (the great majority of the people) and not owned and controlled by a tiny minority of people. Look at the state of the world today! Despite all the wealth and all the resources of the Earth, millions and millions of people in many parts of the world are hungry. Hundreds of thousands of children are dying from hunger every day; and what about the death from diseases, mal-nutrition and lack of medicine and medical help? These are all the consequences of the exploitations (at different levels and in various ways) of human beings by human beings. It is the unjust systems of economy as well as the wicked political systems run by greedy and ruthlessly selfish individuals that give rise to all the injustices and wicked inequalities in the world. Yes indeed, it is a world in which the money and the wealth is in the hands of a small minority instead of being distributed among the people (and for the well-being of the people) of the world.

What troubled Marx, perhaps most, as a result of his societal studies and socio-economic analyses, was his realization, even his total conviction, that the exploitation of man by man, and the injustices also the terrible inequalities that were (and of course still are) so prevalent in capitalist societies and throughout the world are the result of the economic systems of capitalism. Marx wrote a few books and a number of articles and theses. His major work is das Kapital in three volumes; the first volume was published in 1867, the second and third were published posthumously (after having been edited by his friend, Engels) in 1884 and 1895 respectively. The Communist Manifesto was a joint effort, written by both. Marx's struggle was aimed at ending the exploitation of the great majority of the people by a tiny minority of Capitalists who were and are empowered by the wicked economic system called Capitalism. He was not religious nor was he spiritual as such; nevertheless in him there was humanity.

Indeed, in Marx's heart and mind there was some 'Hegelian humaneness'. He better than anyone else knew that Hegel's philosophy and especially his dialectical method has a mystical 'flavour' to it; nevertheless he read it, and he read it diligently and learned from it a great deal. He found it both interesting and rational; he then used it in his philosophical works. Here he is, in a letter to his friend Engels, talking about Hegel's logic - the dialectical method. Let us listen to Karl Marx himself and judge for ourselves, from the language that he uses in this letter, whether he was a hard-headed materialist (as is the common belief in the West) or a human being with some human heart.

He says,

'In the method of treatment, the fact that by mere accident I again glanced through Hegel's Logic has been of great service to me ..... If there should ever be time for such work again, I would greatly like to make accessible to the ordinary human intelligence, in two or three printer's sheets what is rational in the method which Hegel discovered but at the same time enveloped in mysticism.'

On the tombstone of Karl Marx are engraved a few words from one of his theses:

'Philosophers have interpreted the world in various ways; the point, however, is to change it.'

Not only Karl Marx, but many other great men and women of the past, as has been said earlier, were influenced by Hegel ---- learning from his philosophy and his dialectical method. These 'Hegelian relationships' have not always been simple and straightforward. In fact in some cases they have been strange even at times bizarre 'love-and-hate' relationships. One such case is evidently that of

Kierkegaard's antagonistic attitude to Hegel and his philosophy. But before delving into 'Kierkegaardian affairs' with Hegel, I want to say a few words about our own Bertrand Russell's relationship with Hegel.

Bertrand Russell, arguably the greatest philosopher of the 20th century, described Hegel as one of the hardest philosophers and Hegel's philosophy as probably the hardest to understand. Despite experiencing such philosophic difficulties, Russell spent his 'philosophic honeymoon' with Hegel and Hegelian philosophy. This was when he was in his 20s. This Hegelian flirtation and dialectical infatuation lasted several years. Then his Hegelian love began to diminish so that gradually he realized that his love for Hegel is being replaced with a dislike for Hegel and his philosophy. In short, the time came that Russell found himself to be a true anti-Hegelian, which of course is a 'norm' among 'Hegelian Lovers'. Russell, however, does not accept the full responsibility for such 'act of philosophic unfaithfulness' (or, 'Hegelian treachery'). According to Russell, the philosopher **Moore,** and he (Russell) together rebelled against Immanuel Kant and Hegel, and even between them it was Moore who was the 'real rebel' ---- *"Moore led the way, but I closely followed in his footsteps".* There is a book written on this idealist period of Russell's life; its title is: 'Russell's Idealist Apprenticeship' - an apt title too.

Now let us talk of Kierkegaard. Kierkegaard, the Danish philosopher and renowned for being the 'father of existentialism' (and one of my favourite social and philosophic writers) built his reputation, at least partly, at the expense of his 'apparent hatred' to Hegel. Put another way, the 'impetus' for his being a copious writer, to some degree, was his 'obsessive resentment' against Hegel. Many have said that such hatred or resentment was the other side of the 'coin of love'. Whatever it was, one thing is clear and that is

Kierkegaard was an avid reader of Hegel, and was greatly influenced (whether out of love or hate) by Hegel.

Kierkegaard could not be 'the baby of existentialism', let alone the 'father of existentialism' hadn't it been for Hegel and Hegelian philosophy, especially his 'Phenomenology of Mind' (which is usually known as Hegel's greatest work, and sometimes is referred to as the 'Hegelian psychology'); it is notoriously hard to understand. Anyhow, the 'seeds of existentialism', as we know it today, were discretely cultivated in the philosophy of Hegel by Hegel. Put another way, Existentialism is 'embedded' in Hegelian philosophic system, which is the *Department Store of Philosophy*' containing religious, aesthetic, moral, legal, social, political, economic, existential and metaphysical philosophies. Hegel was not just a philosopher but also a philosophic creator ---- unique in this respect. Hegel's philosophical work, 'Hegelianism' as a philosophic system, has had a widespread influence on the Western civilization --- politically, culturally, intellectually, metaphysically and so on. His creative philosophic system influenced greatly (and some would say played important part in) the development of Marxism, Analytic Philosophy, Existentialism, and Positivism.

Hegel wrote many books and numerous articles. Four of his well-known and important books are:

1) The Science of Logic
2) The Philosophy of Right
3) The Encyclopaedia of Philosophy
4) The Phenomenology of Mind

Between 'The Science of Logic' and 'The Phenomenology of Mind' there has always been a 'rivalry'; some say Hegel's most important work is the former, and there are those who say it is the latter.

Hegelianism also had its influence, at least to some extent, on the shaping of the early thinking of **Nietzsche,** the great German philosopher. It is not so difficult to detect in the writings of Nietzsche some of the Hegelian political, religious and existentialist thinking.

Two of the five persons who had their literary, philosophic and intellectual influences on the philosopher **Jean Paul Sartre** were Hegel and Karl Marx. Hegel's philosophy had a great influence on Sartre and his philosophic writings. Many have said that Sartre's 'Being and Nothingness', which is his greatest work, is a 'lengthy footnote' to Hegel's 'Phenomenology of Mind (Spirit)'. I don't know about being a lengthy footnote, but one thing that I know is that Sartre skilfully drew upon Hegel's philosophy, particularly the 'Phenomenology of Mind' for writing his 'Being and Nothingness'. Of course there are many other ideas in his book that are, more or less, original and the product of his (Sartre's) brilliant mind. His 'Being and Nothingness' is a good read and many parts of it are very interesting. Sartre, a true existentialist, was a great thinker and an able philosophic writer, if not a philosopher in the conventional sense.

One of the great men of the 20th century, a pacifist, Christian and visionary, when discussing philosophy with friends, used to say: 'My favourite philosopher is Hegel'; he was a champion of human rights. He was **Doctor Martin Luther King**. Hegel's philosophy played a part in Doctor King's perception of the world, and Hegelian dialectical thinking greatly influenced his thinking. Incidentally, Doctor Martin Luther King was not the only 'man of God' who was so interested in Hegel's philosophy; there have always been spiritual and religious people including some theologians who have shown a great interest in Hegelianism in general and in Hegel's dialectics in particular.

It is not only great people with great minds become attracted to Hegel's philosophy and influenced by Hegelian thinking and Hegel's dialectics. This can also happen or apply to ordinary people like me; I am an unashamed Hegelian. A Hegelian is a person who has some general knowledge of Hegelian philosophic system, and a reasonable understanding of Hegel's dialectical idealism and in particular he or she has adhered to Hegel's dialectics (that is, has accepted the dialectical process as a 'rational method' of the *workings of the things* in this world and even as an 'ongoing process' in the universe).

Hegel's philosophy is very much related to (is concerned with) human mind; that is why there is such a 'thing' as *Hegelian Thinking*, which is a special 'mode of thinking'. Therefore we can see why many Hegelian scholars and students refer to Hegel's *Phenomenology of Mind/Spirit* as 'Hegelian Psychology'. As a Psychologist and a Hegelian pupil, and especially a student of Hegel's dialectics, I see Hegelianism (the whole philosophic system of Hegel) mainly, but not only, as the 'Philosophy of Mind'. I also see it as 'Psycho-spiritual Philosophy'; the spiritual aspect of it is of mystical nature. In Hegel's philosophy I see three modes of thinking in hierarchical order: Holistic, Dialectical and Mystical. What makes them different from one another is chiefly *Perception* (the ability to recognize, the capability of detecting and the capacity for becoming/being aware). We may think of them as three 'types' of mind. I am going to describe them very briefly, as I perceive them. This is my subjective view of certain aspects of Hegelian philosophical psychology. This is not Hegel's psychology according to his *Phenomenology of Mind* in which the three sets of mind are the subjective, the objective, and the Absolute. Here I am going to describe just in a few words the three modes of Mind as I see in and through Hegelian philosophy. These are, as I just said, *Holistic, Dialectical, Mystical.*

A **Holistic Mind** is at the stage of 'wholeness' (special kind of 'maturity'); it sees Reality only in the 'Whole'. It is noteworthy to mention that the person with a 'holistic mind' may not be conscious of his or her 'state of wholism' or even may not have heard of the term *holistic*, yet this does not discard the fact, or diminish the significance, of his or her being a person with *wholistic mind*. Such Mind, as I just said, sees the 'Reality' of a thing in the Wholeness of that thing. Here holistic mind stops and does not go further, but the dialectical mind does; it goes much further.

A **Dialectical Mind** is at the stage of the *'Development of the Self'*; it sees the world and perceives the working of the world, as an exposition of resolving the conflicts or antagonisms, through a series of (conceptual) unfoldings with the aim of attaining to the 'wholeness', thus uncovering the 'reality', for according to Hegel, *reality* is only in the 'Whole'. A dialectical mind is a 'radical mind', and potentially is revolutionary. A dialectical mind is necessarily 'holistic'; and it subsumes holistic mind.

A **Mystical Mind'** is at the stage (or, in the state) of the Growth of The Soul/Self into the Spiritual Realm; it necessarily includes dialectical mind; but then it goes beyond the realm of dialectical mind. A mystical mind sees the complete transformation, 'the revolution', of the mind only in being 'one' (in union) with 'Ultimate Reality'. Though gentle, meek, and even in great majority of instances truly humble, a mystical mind is also revolutionary. Sadly the words 'revolution' and 'revolutionary' are considered by most people to be somehow 'threatening', thus, a revolutionary person is regarded as somewhat subversive. Let me, therefore, astonish these people: The 'desire for revolution', which is the 'desire for complete transformation', stems from a Higher Dimension ('Spiritual Dimension') in a person, and not from the basic instinct of aggression in human being. Yes, yes, this is true. My studies in such areas taught me that the higher is the spiritual development (especially of mystical nature) in a person the more intense becomes the desire to see complete transformation where such transformation is needed.

And Finally, Hegelians (most of them anyway) like to believe that Hegelianism is original and Hegel's philosophy is not derived from any other philosophies. This, more or less, means Hegel was not influenced philosophically by any philosopher. At the same time some Hegelian scholars say that Hegel was (very much) influenced by his predecessor, Emanuel Kant. Perhaps as a young man, Hegel intellectually was to some degree influenced by Kant, whose fierce critic, incidentally, was Hegel himself. (When Kant died in 1804, Hegel was thirty-four years old.) To my understanding, neither a special philosophy had a great influence, nor a particular philosopher had a noticeable impact on Hegel. It was, however, *mysticism* in general, and a half a dozen or so *mystics* in particular, who had a lot of influence and even considerable impact on him. Most Hegelians and many scholars on Hegel's philosophy, sadly, don't know, or don't bother to know about this highly important aspect of the philosopher, Hegel. This 'lazy (or perhaps somewhat arrogant) attitude' of some Hegelians has made their lives doubly difficult as far as the understanding of Hegel's philosophy is concerned. I say this because after all it was these mystical influences, hermetic effects and gnostic impact that gave rise to his *idealist dialectics* central to which is the concept of "Wholeness and Reality" (that 'Whole', and only *the whole*, is 'Reality'); this is an 'ascending progression' leading to or ultimately culminating in the attainment to what Hegel calls *Absolute Spirit*.

Hegel's interest in *mysticism* (mystical spirituality), which is said to be the highest form of spirituality, began when he was very young ------ perhaps about fourteen years of age. When he was in his mid-twenties, he was well into mystical spirituality (many say hermetic spirituality and there are some who use the term, 'gnostic spirituality'). The three have only in general some similarities; they are different 'Spiritual Ways' of Life and have different approaches to the attainment to *Ultimate Reality*. Hegel had read and studied diligently plenty of literature on mysticism. Two mystics that impressed him intellectually most and influenced him spiritually more than others were Meister Eckhart and Jacob Boehme. Some have said (including his contemporary philosopher, Schelling) that Hegel's philosophic work is largely based on Boehme's mystical views and esoteric knowledge. Yet and yet, even to this day, many followers of Hegel dispute his spirituality, and even still the argument among some Hegelian scholars goes on whether Hegel was a Christian or an atheist. Isn't it amazing?

Of course it is amazing. Here is a man whose 'core of philosophy' is *mystical* and also the center of his philosophic system is *Absolute Spirit*, yet he appears to most of his readers to be 'not-so-spiritual' man; even to some he sounds more of an 'atheist'. I can see why such confusion exists. Hegel's mystical spirituality is clothed with his philosophic creativity, which in turn is covered with his intellectual complexity. Hegel, with the exception of Karl Marx (but then Marx himself was profoundly influenced by him), has been, particularly in the West, the most influential philosopher ----- politically, socially, and intellectually; and his certain philosophical concepts and insights were influential in a number of fields and disciplines of science. To Hegel the job of philosophy is to expose the truth and to uncover the reality. And throughout his philosophic system Hegel endeavoured to do this.

The late Iris Murdoch, that brilliant novelist, who was also a philosopher, had a deep insight into Hegel and Hegelianism. She was, in my opinion, a Platonist in mind and a Hegelian at heart. I have enjoyed tremendously reading some of her philosophic writings --- and of course I have also learned from them. I remember in one of her books she says, **'In the philosophy of Hegel there is more truth than in that of any other philosopher.'** Absolutely true. And it is this 'truth' (Hegelian truth, of mystical nature) combined with the transformative aspect of his dialectics that has throughout attracted certain minds including some of the greatest ones, all dialectically seeking the truth.

Addendum

The little knowledge that I have of Hegel and my little understanding of Hegelian philosophy began with 'probably the smallest' book written on Hegel. This 'great little book' is written by Professor Peter Singer: it is, in my view, the best Introduction to Hegel and his philosophy. I recommend the book in question to anyone, especially the beginner, who wants to acquire some knowlege of Hegel and his philosophy. Incidentally, it is from this book that I have quoted in my essay the letter of Karl Marx to his friend and collaborator, Engels; and I have done so without prior permission. I know and I am sure Professor Singer has already forgiven me. Thank you Professor for writing so succinctly and with such clarity on so abstruse philosophy in a very brief book----thank you.

6

Mystical Psychology
And
The Revolution of Mind

Mysticism or 'mystical spirituality' is known to be the highest form of spirituality. Mysticism is a spiritual way of life, which essentially includes some systematic practices and also certain disciplines some of which are often of ascetic nature; its final aim is 'Union with Ultimate Reality'. On the Mystical Path total commitment, dedication and perseverance are essential. These practices and disciplines vary in kind and in degree among the mystical aspirants, depending on the belief system of the person on the 'Path'.

The aim of mystical aspirant, who is on the Path, is to get closer and closer to and finally to become 'one' with a Sacred Entity, Divinity, or with Ultimate Reality, God. Such 'spiritually noble ambition' usually takes years, for some individuals one or two decades and for some it can be a life-long pursuit in order to reach the *Final Mystical State*. And for some people such 'mystical ambition' may not be fully attainable. Different religions and belief systems refer to the 'final mystical state' by different names.

To be a committed 'traveller' on the Path is not all rosy, nor is it full of mystical blessings. It is true that 'on the hills and in the valleys' of the mystical journey there are spiritual joys, and some blissful moments and even every now and then there may be among some advanced aspirants certain glimpses of *mystical eternity*, but there are also mental anguish, emotional turmoils, spiritual agonies, existential emptiness and existentialist 'dread and despair' (angst); and what about going through the 'Dark Nights of the Soul'?

Mystical schools of different belief systems have different explanations for such psychospiritual anguish on the spiritual path especially of mystical nature. One explanation is that the mind on the mystical journey is resisting to yield to the Spirit, hence, the conflict and tension in the individual's psyche. In Christianity it is the 'old creation' that is not giving way to the 'new creation' (the phrase used by Apostle Paul). Sufis say that the 'lower self' is adamant about going through 'Fana' (non-existence, annihilation), in order to pave the way for the 'Higher Self'. Indeed, the destructive elements of personality such as pride, aggression, jealousy, revenge, lust, hatred, greed, arrogance etc. cling to the 'negative compartment' of their lower self (the old creation) and do not easily give way to the 'noble aspects' of the Higher Self (New Creation), such as forgiveness, compassion, loving kindness, generosity, gentleness, and humility. However, on the Spiritual Path, through perseverance and spiritual obedience, finally the old creation 'surrenders' to the New Creation; or as Sufis would say, the lower self yields to the Higher Self.

Pains and suffering in human life particularly of psycho-spiritual nature can be (and they are for those who want them to be) the best opportunities for self-discovery. To run away from our afflictions is to run away from our 'selves' and from the realities of our lives. Let us also not forget that hidden in our sufferings are many truths about ourselves. It is by accepting our pains and going through our difficulties in life that we can discover such truths about ourselves. Furthermore, confronting our pains courageously and going through our sufferings acceptingly are the best lessons for the growth of the Soul: learning patience and forbearance, obedience and humility. "Son of God though He was, He learned obedience in the School of Suffering." (New Testament, Hebrews).

Spiritual development, especially of mystical nature, is holistic; it embraces the development of the whole person. This means that such development 'touches' the person in all dimensions and at all levels. Spiritual development in particular includes the psychological development of a person. However, psychological development does not include spiritual development, but it may for some people be a 'preparatory ground' for the attainment of spiritual growth. In short, mental growth and spiritual growth are two different faculties, two separate dimensions. We often meet men and women who have attained mental growth and intellectual development, but many of them live a life that is 'spiritual desert'. However, mind is an *avenue* leading to Spirit (of course for those who want to use it as such.)

Those individuals who have succeeded on *The Path of Self-Discovery* have also succeeded in attaining personal growth ---- mental growth and emotional maturity. On the mystical 'ladder' the mind of the mystical aspirant goes through radical changes that are capable of leading the mind, over a period of time, often over many years, to 'complete transformation', a kind of 'psychospiritual revolution'.

Freud, that admirable genius, revealed and described to us the immenseness, even the limitlessness, of the mind. His psychology is *psychoanalytic psychology*, or to put it in one word, 'psychodynamics', which refers to the working of the inner mind. Freud showed us the 'underground' of the mind, the dark valleys and even the frightening narrow passages of the mind. He demonstrated the world of the 'unconscious'------- the deeper layers of the mind. In short, Freud's Psychology is, so to speak, a 'descending psychology', i.e. it is mainly the study of the deeper layers of the mind. What we need now is an *ascending psychology*.

As I just said, n*ow* what we desperately need is an 'ascending psychology', psychology of Higher Levels of Consciousness. Central to this is the study of the 'Self' within mystical realm; this is *'The Science of The Self'*. The key feature to this science is the study of the complete transformation ('revolution') of the mind on the mystical journey. *In truth,* **the 'Science of Self' is the Study of Complete Psycho-Spiritual Transformation through the Higher Levels of Consciousness. This is 'Mystical Psychology'.**

Our departments of psychology, perhaps with the exception of Humanistic Psychology, still feel shy or inhibited to use the word 'Self', let alone making the 'Science of Self' as part of their curriculum. But time will come (it has already begun) that the 'centre of being', Self, will be central to the teaching of psychology at our universities. Science of Self, *mystical psychology*, will in future become an important part (branch) of the western psychology. Such 'science' has been an important part of the psychology of the East; in fact in India the 'Science of Self' has since ancient times been an essential part of its philosophy. At this point I am going to talk briefly about the fundamental differences between the psychologies of the East and the West.

The fundamental difference between the psychologies of the East and the West arise from the fact that in the East, particularly in India, the study and the discovery of the Self is inherent in their psycho-philosophy, central to which are questions such as what is 'normality' and what is 'abnormality' in human being? What is 'mental health' and what is 'mental illness'? Eastern and Western psychologies, have entirely different answers to such questions. According to the Western Psychology, a person who is 'not mentally ill' is normal, therefore, he or she is 'mentally healthy'. (The term, 'mentally ill' is a psychiatric term, referring to persons who are unable to behave in a socially acceptable manner, and also cannot, chiefly due to 'some mental problems', function in personal life responsibly.) The criterion for 'normality' in the West ("not being mentally ill") also can, and very often does, apply to some of the most notorious criminals and dangerous people including gangsters, serial killers and even those who have committed genocides. We all have heard or read in the papers how a serial killer, when arrested after committing many killings over a period of many years, is described by all his friends, colleagues and neighbours as a very nice, decent and well-behaved human being!

The problem with Western Psychology is that it does not distinguish between 'being mentally ill', and 'having a sick mind'. It is very important to make a distinction between having mental illness and being mentally ill. Winston Churchill lived for many years with mental illness (and a severe one) but he was not mentally ill. Some of the great artists and writers in the world have, or have had, mental illness but have not been mentally ill. Adolph Hitler and many of his men had sick minds but they were not mentally ill.

According to the Eastern psychology in general and the Indian psycho-philosophy in particular, human beings are born as 'unfinished' creatures and they will remain so if they do not consciously and actively work on themselves, aiming, on the Path of Self-development, at 'Self-discovery'. We are told these 'unfinished beings', human beings, live in their illusions, ignorance and fears, which progressively give rise to their misperceptions and misconceptions of life and the world that they live in. The only way that individuals can free themselves from their unreal world (the world of delusions, fears, misconceptions and so on) is to make the effort of 'finishing', completing, themselves; thus, by working on themselves they may awaken from their illusional world and delusional life. This, on the 'Ladder of Self-realization', may eventually lead to 'Awakening', *"the Enlightenment"*.

The mind of a person (say, a 'mystical aspirant') that has reached a reasonably high degree of Self-realization, though probably not fully 'revolutionized', it has been through a good deal of transformation. A revolutionized mind (mystically revolutionized) is also a revolutionary mind, but a revolutionary mind (in its conventional or political sense) is not a 'revolutionized' mind. Here, in this essay, by 'revolutionized mind' I mean a mind that has mystically been through 'complete transformation'. It is simply the complete *'Psycho-Spiritual Transformation'*.

Mystics are not, as many people think, reclusive, self-absorbed or indolent people. They are active and full of energy. They are steadfast, resolute and determined individuals. They are gentle, and also caring towards their fellow human beings; they are truly humble. They cannot bear to see social injustices, unfairness, cruelties and the wicked inequalities, which are so prevalent in our societies and in the world that we live in today. Indeed the mystics, enlightened people, are truly decent and really moral beings. The **late Iris Murdoch**, one of the best British novelists, and also a philosopher, says,

> **"The mystic is neither irrational nor a dreamer.**
> **Mystic helps the world in practical ways.**
> **Mysticism is the everpresent moral idea**
> **extending ordinary decent morals indefinitely in**
> **the direction of Perfect Goodness."**

Indeed, "Perfect Goodness". They aim at Perfection or Perfect Goodness. Jesus said, 'Be Perfect, even as your Heavenly Father is Perfect'. Yet, so many of the followers of The Perfection and seekers of The Truth, were throughout the centuries treated shabbily and many of them were cruelly ostracised and even some of them brutally killed. (Here I have in mind the 'Theistic Mystics'.) They were ridiculed and regarded as crazy eccentrics, and shameless heretics by the religious authorities. But why these decent and illumined people were treated so shabbily and so cruelly? The answer is, because of upholding their 'mystical principles'-------refusing to obey the religious authorities and rebelling against man-made dogmas and the absurd doctrines of the religious institutions. Mystics instead of following religions focus all their attention on God, aiming at having direct experience of Him and seeking wholeheartedly His Love, and waiting obediently, prayerfully and meditatively at being 'in Union' with Him, Ultimate Reality.

Anyone who has read a good deal of mystical literature, the writings of the mystics and their poetry will attest the things that I am saying here. In them (in the spiritually awakened people) there is bliss and constant joy, and at the same time one can detect in them a kind of anger, which is *Moral Anger* stemming from their purity of heart. The Man who turned over the table of the greedy money-exchangers in the Temple showed us His Revolutionary (Divinely revolutionary) Nature. And this Man was condemned to death partly because He challenged the ruling authorities. He publicly and fearlessly spoke against some of the unseemly and unholy State of Affairs and also against the hypocrisy that was so prevalent among so many people including many of the rulers and the elders at that time.

Only the Minds that are advanced in Higher Levels of Consciousness become capable of transforming completely the state of affairs, religious, social, political or whatever. Throughout his life,

St Francis of Assisi (a religious mystic and known to many Christians as the 'saint of the saints') knew that in human being there is a deep desire to establish a relationship with God, and he wanted to see a kind of mystical worshipping of God and not a ritual worship that is based on dogmas and man-made doctrines. Let me quote from a book whose title I can't recall (I am sure the author will forgive me).

"St Francis was a man, who deinstitutionalized the medieval Church, represented a mystical Christian tradition and stood up to leaders and all great citadel of monastic influences."

This kind of saintly rebellious nature is not restricted to one particular religion, or certain spirituality; it applies to all mystics of all belief systems. In their mystical literature we read the deep empathy of 'the Awakened' towards the sufferings of their fellow human beings. They are special human beings. And their 'Special-ness' is in the fact that they have been able to revolutionize their Mind and to discover their *Self. M*oreover, they have gone on endeavouring, on the Path of seeking the Truth, *to experience illuminating glimpses of mystical eternity, which they love to share with their fellow human beings and the World that they live in.*

In Mystical Psychology, the Science of Self, the main features are Self-Discovery and Seeking the Truth. Put another way, it is the Study of the Complete Psycho-Spiritual Transformation of Human being chiefly through attaining the Higher Levels of Consciousness. Such science is, without any doubt, a 'Noble Science' (in my view the noblest). The Science of Self, Mystical Psychology, which is an ennobling field of study, ought to be an important branch of the field of psychology in all the universities. Let us hope this 'Noble Dream' will materialize in near future and will revolutionize the minds,

7

Simone Weil

Simone Weil was born in a French Jewish intellectual and fairly wealthy family in 1909 in Paris, and she died in 1943 in Ashford, Kent, England; she was 34 when she died. She received the best education. She studied philosophy and classical philology, specializing in Greek philosophy; she obtained her Doctorate in philosophy. Furthermore, she knew several European languages and a couple of Eastern languages including Sanskrit.

Simone Weil was amazingly precocious. Her extraordinary intellectual capacity was manifest while still a child; and it was combined with a sense of social concern for her fellow human beings anywhere in the world. At the age of five she refused to take sugar when she heard French soldiers did not have sugar at the Front in First World War. The news of the starvation, killings, or any social injustice in any part of the world would distress her; and often so much so that she was prepared (even at the age of ten or twelve) to go and help the victims.

For a while she taught philosophy in a girls' school in Paris. She also edited one or two Journals of Left-wing affiliation. She did not join the Communist Party because she saw the Communism of Soviet Union had nothing to do with Socialism. She detested greed and materialism and deplored the exploitation of man by man for the sake of profit and getting richer and richer. In order to know about the workers' conditions in the industrial places she joined a car factory, Renault, as a worker on the production line she was appalled by some of the appalling conditions and the exploitative environment that her fellow workers were working in. She also worked for short periods in two other factories. Later she wrote about the periods of her 'industrial experience'.

She was about twenty-seven when she left for Spain and there she joined the Anarchist Unit of Republican Force in Spanish Civil War. Then, when she came back, for a while she worked in a countryside farm, as a farm servant.

Simone Weil (perhaps from the day she was born) was not an ordinary person. She was born with a 'revolutionary' mind (perhaps 'revolutionized' is a better word in this context). And this brings me to another aspect of the 'personhood' of this extraordinary woman. The 'aspect' that I have in mind is the 'dormant spirituality' in her. In Simone Weil, a non-believer, there was a potential spirituality.

Though she was born in an agnostic family, a completely secular ambience (her parents were intellectual agnostics), Simone manifested a great capacity for 'spirituality', as later we shall see. In my view underlying her deep concern for her fellow human beings and her loving care for those who were in pain and suffering and also her genuine desire to bring about radical changes where social injustice existed was the mind of a 'spiritually-awakened' person, yet not on the conscious level. But before going into the dormant (potential) spirituality of this 'non-believer' called Simone Weil, it is necessary, even essential, to talk, at least in brief, about her extraordinary intellectual capacity to which I have already, in passing, made a reference.

The intellectual capacity of Simone Weil was more than extraordinary; it was truly amazing. The impact of her intellect and her social views, moral concepts and philosophical ideas began in her early 20s. Two of her close friends were the Philosopher Jean Paul Sartre and his friend and companion, the renowned intellectual, Simone de Beauvoir. Albert Camus, the French Philosopher said of Simone Weil, "The greatest social thinker since Karl Marx". The Pope, (Paul VI) said that Simone Weil was one of the three individuals who had the most important influence on his intellectual development. (The other two were the philosopher and mathematician, Pascal, and Bernanos). Her professor of philosophy, Alain (a pseudonym), said, "She had a power of thought that is rare". She had dialogues and sometimes heated debates with the famous Russian revolutionary, Trotsky, on Marxism, politics, Hegelian dialectics, and other philosophical issues; he was impressed by her intellect and power of debate.

The late Iris Murdoch, the well-known British novelist and also a philosopher, was a great admirer of Simone Weil. Simone was one of the few people who had intellectual influence on Iris. Weil is one of the most mentioned names in Murdoch's philosophical writings.

Maurice Schaumann, the distinguished French politician, who later became France's Secretary of State for Foreign Affairs, was a close friend of Simone. As students of philosophy in the classroom they sat next to each other, listening to Alain's philosophical lectures. (Alain's real name was Professor Emile Chartier.)

Charles de Gaulle was the only person who did not like Simone Weil; in fact he could not stand her. He referred to her as 'That mad woman'.

The spiritual aspect of 'the fully secular' Simone Weil, it seems, manifested 'suddenly'. Such 'suddenness' (more of a mystical kind) took place when she was on holiday in Italy in 1937. However, before quoting her description of her spiritual experience it is important to say a few words about Simone's attitude to Christianity and to Christian believers. And let us see how things change when God wants them to change---it is astonishing!

Not only did Simone not have much opinion of Christianity, but also she thought of it as nothing more than a superstition. She could never understand why people kneel down in front of the Cross, crucified Christ; in fact she derided them. Now let us listen to her words, describing the spiritual experience that I just mentioned. She says,

"In 1937 I had two marvellous days at Assisi. There, alone in the little 12th Century Romanesque chapel of Santa Maria degli Angel an incomparable marvel of purity, where Saint Francis often used to pray, something stronger than I was compelled me for the first time in my life to go down on my knees."

And she did go down on her knees at the foot of Cross and for the first time in her life she prayed. (Not long ago I read somewhere that she experienced this 'spiritual phenomenon' while she was looking at the Crucified Christ ------ something that she always derided!) It was then, as she puts it, 'Christ took possession of me'. If this is not a *Wonder*, then I wonder as I wander what is a 'Wonder'.

Then after that spiritual experience she began to read the Bible and also to listen to Spiritual Music and especially to read metaphysical books, spiritual literature and mystical poetry. She fell in love with the poem of 'Love' by the English metaphysical poet,

George Herbert (1593-1633). She used to recite it again and again everyday. In a letter to a close friend who had become permanently paralysed, she says,

"I enclose the English poem, Love, which I recited
to you. It has played a big role in my life, because
I was repeating it to myself at the moment Christ
came to take possession of me for the first time.
I thought I was only reciting a beautiful poem,
unknown to me it was a prayer."

Later, in the space of two or three years, Simone on several occasions encountered, went through, several mystical experiences. She also found an affinity with the Roman Catholic Church. She would go to church though not regularly. She also took Communion. However, She did not join the Church, i.e. she did not become a member. This was on the ground of not agreeing to certain 'Doctrinal Issues' of the Church. This (not becoming a member of religious institutions) is not uncommon among the mystics and also often among mystically-minded individuals. This is because mysticism (mystical spirituality) and conventional religions, hierarchical religious institutions, are not compatible and do not easily get along with each other. The truth is that mystics do not feel at home in religious institutions. They feel at home only when they are at 'Home'-----where they belong.

Weil's political views and social values gradually became 'spiritualized'. It saddened her deeply to see the Western capitalism with all its capitalist societies becoming more and more materialistic and loosing their spiritual values ----- too much preoccupation with materialistic possession. And at this point, as I am writing these lines, I say to myself,

"Oh Dear Simone,

that was sixty years ago; come and see now what is happening!
Come and see the expansion of ruthless capitalism. Now we have 'Global

Economy and Global Market' which have given rise to Global Greed. Indeed, money-mindedness and greed have become the worst kind of disease. It is all a Ruthless Rat Race! At the same time we have millions and millions in many parts of the world starving to death. Big countries are competing with one another in building and selling Weapons of Mass Destruction to any country or any dictator who wants to buy them. Billions and billions every year are being spent on weapons of war but not spending half, even a fraction, of that money on 'bread and butter' to put in the mouths of the hungry children of the world. The morality that you were talking about is hardly left, and the moral values that you were dreaming of never materialized. Societal decency and Human values are all going so rapidly down the drain.

Simone, don't come back. Stay where you are and enjoy the 'Mystical Eternity', I am sure, where you are phrases or statements such as 'Global Economy', 'Multi-National Companies' are non-existent."

Simone Weil saw clearly the unfairness that went on around her, and she perceived vividly the social and political injustice that was going on not only in Europe but also throughout the world. She was well aware of the exploitation of human beings by their fellow human beings for the sake of profit, becoming ruthlessly richer and richer. And after all she also experienced it when she worked in Car Factories. She says,

> **"Modern industrial organizations in particular
> have an exploitative capitalist structure which
> Puts profit and production before human beings,
> and this depersonalizes and dehumanizes them."**

Simone saw a need for social rootedness, but not as something separate from the Spiritual rootedness. She says,

> **"There is a need for people to be rooted in
> cohesive social groups to which they belong,
> and also there is a need for being rooted in
> Spiritual realm."**

She was a 'born Socialist' (if there is such a thing). Her Socialism was an ethical one, which is firmly rooted in the Teaching of Jesus. This kind of Socialism is embraced throughout the world by millions and millions of people from all walks of life and from all social, cultural, racial, and religious backgrounds. Socialism must, first and foremost, have a 'Moral Foundation' and not be based on 'ideological indoctrination'. This kind of socialism stems largely from the spiritual dimension that is inherent in every human being whether one is a true believer or a militant atheist. Having said this, I must also emphasise that there is nothing wrong if such socialism is also influenced by some 'Marxian thoughts'; in fact this may enhance, intellectually, the socialist conviction of many socialists.

Here it is noteworthy to examine briefly certain important aspects of the spiritual beliefs and concepts held by Simone Weil. And I will start with her view on the spiritual growth in a person, especially how to maximize such development of the spirit. Some writers refer to the spiritual views and metaphysical concepts put forward by Simon Weil as 'the theology of Simone Weil', or Weil's 'mystical theology'. The term 'mystical theology' strictly speaking is a contradiction in terms. Anyhow.

According to Simone Weil in order for our spiritual growth to reach its completion 'the loss of the self' is essential. She describes this as 'decreation'----- to decreate the 'old self' and to recreate a 'new self'. In fact what she says is in complete agreement with the saying of Jesus: 'if you don't loose your life [old self], you cannot gain your life [New Self]. Likewise, Sufis say, "If you don't annihilate your *lower self*, you cannot attain to your *Higher Self*".

Weil also along with her concept of 'decreation' introduces the practice of 'attentiveness'--------the mind to be constantly attentive to God. This is very similar to 'one-pointed concentration' in Yoga. Such concentration, which is also referred to as 'meditative concentration', is an essential part of yogic discipline.

There is, however, 'something' which is probably more important in Weil's Mystical Discipline. This is in my opinion fundamental to her teaching and to her life. This is her concept of human sufferings. While accepting that suffering is horrible and no one in one's right mind wants to experience pain, she says when we encounter our afflictions we must not run away from them but go through them. She says to run away from our sufferings is to run away from the realities about our 'selves'. She goes on saying that a lot of truths about us are 'hidden' in our sufferings. The afflictions, pains and sufferings, of every person she says, 'contain' important truths about that person. It is only by facing them acceptingly and going through them obediently (she uses the word 'obediently') that the persons can gradually know, and have some understanding of 'the truths' about themselves. This means our understanding of our suffering gives rise to our understanding of our 'Self'. However, Weil does not stop here; she goes much further. She says that through understanding our afflictions we can also 'know something about God'. Indeed! I fully concur with Simone on the concept of human suffering and its relation to our personal growth and Self-knowledge. The mystics of all belief systems say that Self-realization is the road leading to God-realization; in fact the two are closely related.

For Simone *Human Affliction* is the reminder of *the Cross of The Christ*. She does not say this for the sake of saying it. She speaks mainly from her life experience. She arrived at such conclusions largely experientially. She discovered them for herself. Her psycho-spiritual scars were deep and painful. She also had physical pains including severe headaches. But her pains and sufferings brought her closer and closer to God and enabled her to know Christ and to understand the meaning of the Cross of Christ. She says,

> **"The Cross of Christ is the only Source of Light that is bright enough to illuminate our afflictions. That is why Cross is our only hope ......Affliction contains the truth about our condition. They alone**

**will see God who prefer to recognize the truth and
die, instead of living a long and happy life, but in
a state of illusion."**

It was 2ⁿᵈ World War with its Nazism in Europe. Life in
France as in most European countries was becoming difficult
particularly for Jewish people. Weil family, a Jewish family, had
to leave France. They left for New York in April 1942. From there
Simone wrote a few letters to Maurice Schaumann. She begins her
first letter by reminding him of the days that they used to listen to
Alain's philosophical lectures. Then she comes to the serious part
of the letter in which she is asking Schaumann to facilitate the
process of getting her a Visa so that she can come to London in
order to join The French Resistance Movement. She also encloses
a detailed plan for organizing 'Front-Line Nurses' in Free-France
Movement. In her following letters she urges Schaumann to
act quickly in getting her a Visa ---- she couldn't wait. In these
letters she also writes of her personal life especially of her spiritual
progress. For example it was in her third letter that she included
the following statement:

**"I adhere totally to Christian Faith with the kind
of adherence which seems to me--------. That
adherence is Love,----. I certainly belong to Christ,
or so I hope and believe."**

Finally, Simone arrived in Liverpool, England, in *November 04
1942*, and then she came to London. She joined The Resistance
Movement and became actively involved in it for freeing France
from German occupation. She identified herself with the people of
France by refusing to eat more than the official ration given to a
French person in France. Soon her health deteriorated. She later got
tuberculosis. While in sanatorium she died on 24ᵗʰ August 1943 in
Ashford, Kent-----a lonely death at the age of 34.

In her funeral there were only seven people including Maurice
Schaumann. The priest, a good friend of hers, could not come due

to some accident. Maurice read a couple of passages from the Bible. Who on earth could predict, when Simone and Maurice used to sit next to each other listening to Professor Alain, that one of them one day would 'officiate' the funeral of the other? Sad world indeed! Life is unpredictable and nothing is certain.

Simone wrote a lot but not a book. Her writings were mostly notebooks; there were also many essays. Her friends and associates collected these writings and posthumously published them as books. These books are read not only in Europe but also in many other parts of the world. Simone's writings are mainly philosophical and spiritual; many of them are of mystical nature and some of them are spiritually moving. The three of her books that perhaps have the largest number of readers are: *Gravity and Grace'*, *'Waiting for God'*, and *'The Need for Roots' (L'Enracinement)*.

Once a secular political activist, Simone Weil became a deeply mystical thinker. But are these really the right descriptions of Weil. Many books including some Encyclopaedias describe her as 'mystic-philosopher' or 'mystic and philosopher'. In some books she is referred to as 'mystic and religious writer'. Some of them describe her as 'The greatest mystic-philosopher of the 20th century'. It is really difficult to classify Simone ----- Political Activist, Anarchist, Mystical thinker, Philosopher, Spiritual Anarchist, Neurotic Saint; or a Revolutionary? Which one was she? Or, was she all of them? Or...What?

The main difficulty, in my view, in classifying Simone is that she was not an ordinary person, not a normal person in its conventional sense. Simone was a 'peculiar' person. Her mind was not complicated, but was a *Highly Complex Mind,* and in a way also an 'unsettled mind'. Yet this highly complex mind was in many respects a child-like mind. I see Simone as a person with a highly developed mind and extraordinary intellect seeking the truth in the simplicities of life-----in a child-like manner. Paradox? Indeed. She was a creature of 'paradoxes' but not of contradictions. This intellectually extraordinary human being was searching for 'Truth' in ordinary and simple things in life ---- a life so absurd and often cruelly absurd. Yes, even in the absurdities and senselessness of life she was in the search of the Truth.

With the academic qualifications and the brilliance of intellect that she had she could become an eminent academic, most probably as a professor in the department of philosophy of a first rate university, and she could have a luxurious life in the most beautiful parts of France or anywhere in the world for that matter. But she did not want that. She wanted what she wanted. She could not stand the social unfairness and political injustices; the wicked inequalities of this world tormented her. She shrewdly observed the absurdities of life and perceived the senseless cruelties, incomprehensible afflictions and the horrors of life that are so prevalent in this world. She was endeavouring to do something about them, or at least to make some sense of the senselessness of life. She tried them all: Anarchist, Political Activist, Freedom Fighter, Factory Worker, Farm Servant, Revolutionary and so on. And then she turned to Spirituality, *Mystical Spirituality*, which led her to rest in the 'Bosom of Christ' where she wanted (perhaps since she was born?) to be.

When Simone Weil died her Professor of Philosophy said of her,

"She had a power of thought that is rare."

And I say of her,

"She had a Humanity that is rare."

There are several biographies of Simone Weil, and in my opinion the best one is written by Professor David Mclellan who is a political thinker. The title of the book, *Simone Weil: Pessimist Utopian.* In the conclusion of the book he says,

> **'Nevertheless Weil remains peculiar.**
> **This French version of Kafka, this cross**
> **between Pascal and Orwell remains**
> **unclassifiable. She is intellectually stateless,**
> **a prophet without any country in which she**
> **can be sure of honour.'**

I am going to end this essay with Simone's favourite poem of 'Love' by the well-known English poet, George Herbert (1593-1633). Simone used to recite it again and again and every day. One day while reciting it, as she puts it, 'Christ came to take possession of me for the first time'. Here it is:

Love

(By George Herbert)

Love bade me welcome; yet my soul drew back,

Guilty of dust and sin.

But quick-eyed Love, observing me slack

From my first entrance in,

Drew nearer to me, sweetly questioning

If I lacked anything.

"A guest", I answered "worthy to be here";

Love said, "you shall be he".

"I, the unkind, ungrateful? Ah, my dear,

I cannot look on Thee."

Love took my hand, and smiling did reply

"Who made the eyes but I?"

"Truth, Lord; but I have marred them: let my shame

Go where it doth deserve."

"And know you not", says Love "who bore the blame?"

"My dear, then I will serve,"

"You must sit down," says Love "and taste my meat.."

So I did sit and eat.

8

Energy Medicine:
Einstein's and Newton's
Models of Reality

There are, apart from orthodox medicine, numerous approaches to the restoration of health and to bringing about healing in human beings. These are usually known as alternative or complementary medicine, and also sometimes are called unorthodox medicine. These approaches are different from one another in the methods and techniques and in the ways that they work in effecting healing and in bringing about a sense of well being in the patients. However, despite all these differences they share one common belief; in truth, it is not just a belief but also a 'fact', which sadly Western traditional medicine as yet has not acknowledged. The belief (fact) in question is that in human being there are patterns of energy usually referred to as bioenergy or 'subtle energies', which continuously flow through their proper channels or pathways to all parts of the body, nourishing it even at cellular level. These energies in human being and also in any other living creature are the product of life (the state of being alive); energy emanates in and from a living being, and is measurable as an electromagnetic field.

In 'Healing Medicine' nearly all, great majority, of the therapeutic approaches come under the 'umbrella' of energy medicine, and also each one of them, individually, may be referred to as 'energy medicine'. These healing approaches are different in their methodologies and applications; they also vary in their therapeutic effectiveness, depending on certain factors. Each approach may be more efficacious for certain ailments, and perhaps also better suited for particular people. For example, acupuncture, homeopathy, herbal medicine and crystal therapy are therapeutically effective for many ailments, yet each one of them could have a greater healing effect than others for certain illnesses, and probably also for particular people.

Energy medicine is a special form of 'art of healing' and a highly important one too. Many approaches of this form of healing are genuinely therapeutic. In every approach the therapist's, or practitioner's, aim is to bring about healing in the patients mainly by stimulating the healing resources, especially by reactivating the bioenergies that are available in them. These approaches are completely different conceptually and methodologically from Western conventional medicine. *The fundamental difference between them is 'conceptual' ---- the Models of reality that they adhere to. These Models are totally different from one another. Here I especially have in mind Einstein's and Newton's Models of reality.*

Newton saw 'matter' as 'solid substance' and nothing more. Einstein saw 'matter' as the 'storage of energy'. To Newton 'mass' was solid object. To Einstein 'mass' is another form of energy, and in fact by his famous equation, $E=mc^2$, he demonstrated that 'mass' and 'energy' not only are two sides of the same reality but also they are convertible. Furthermore, Newton had a mechanistic view, mechanical concept, of the universe. In Newtonian (Classical) physics, the universe is an infinitely huge place containing gigantic isolated objects. In short, Newton's model of reality of the universe is purely mechanistic. Einstein's model of reality of the universe is very much vibrant. For Einstein every object (planet, star or whatever) in the universe is a 'bundle' of energy and *throughout the infinity there is a web-like interconnection of interacting cosmic energies.*

Now we can see clearly the difference between the two models of reality in question. And here it is important to mention that Newton's (classical) physics was the culmination of Descartes' concept of Mind and Body, known as 'Cartesian concept of mind and body'. Descartes saw mind and body as two separate things, functioning independently. He thought of body as a physical object, containing various parts and organs, and obeying physical laws. And such was his view of 'mind and matter' too -- mechanistic.

(it is interesting to see how a philosophical concept, Cartesian concept of mind and matter, through its continuing influence culminated, after a century or so, in the science of Newtonian physics).

Western traditional medicine has always adhered to the Cartesian concept of mind and body, and its model of reality has always been Newtonian - mechanistic. It sees the patient as a physical body only, and patient's illness as physically caused only. It does not believe that certain psychological processes mainly of emotional nature can adversely affect the body and contribute to the development of illnesses or diseases. Nor does it believe that the 'root cause' of many of the illnesses, at least in part, is energy-related, that is due to the abnormal functioning of the energies in the body. Western orthodox medicine sees human body, as Descartes saw, as a physical object obeying certain physical laws. Of course this is not true. Human body is a 'living thing' in which there is the continuous flow of bioenergies, subtle energy. And human body does not obey physical laws, but the laws of 'multi-dimensionality of human being' mostly of psycho-physiologic (psychosomatic) and also psychobioenergetic nature.

Is it any wonder then why Western conventional medicine is not, and has never been, a 'healing medicine'? All its advances are technological. It can perform all kinds of surgical operations and carry out certain organ transplants and often with great success, yet, it is helpless in dealing with many ill-health problems, and has no clue about the causes of some of the serious diseases, which are the biggest killers, let alone to do something positive about them. And the main reason for this is that the western conventional medicine is purely a 'physical medicine' and not a 'Healing medicine'. It can deal fairly successfully with illnesses whose causes are physical or purely biological, but it is helpless in doing something about illnesses or diseases that are multi-causal. And the greater is the 'multifactorialness' of a disease, the greater becomes the helplessness of the conventional medicine in doing something positive about it. Perhaps the best example would be 'Cancer' the research into which has become an 'industry'------ a shameless industry. What is needed now is a 'Healing Medicine'. This means a medicine that sees human being as a multi-dimensional being, and believes that the causes of most of human illnesses and diseases are not physical only but, in varying degrees, are multi-factorial (multi-causal). Such medicine will take into consideration the 'healing resources' of human being -------- including and especially the bioenergies.

So far I have discussed briefly Energy Medicine and also I have talked about Newtonian and Einsteinian Models of Reality. Furthermore, we have seen how well Einstein's Model of Reality and the concept of Energy Medicine go together-----they fit in very well. Now I am going to talk briefly about human bioenergies, human subtle energy. Human energy system consists of two systems of energy. One of them is 'Indian Chakra System' and the other one is 'Chinese Acupuncture Meridian System'. Though two entirely separate systems, Chakras and Meridians are not unrelated; in fact those who specialize in bioenergetics tell us that the two systems are associated with each other and functionally there is a 'working relation' between them.

According to Indian Chakra System there are seven major energy centres starting from the base of the spine and ascending along the spine to the head. These centres, known as Chakras, are not the generators of energy but the 'transformers' of energy - transforming the 'raw energy' into 'usable energy'. There are also hundreds of minor Chakras located in different parts of the body. Each one of the major Chakras 'governs' a particular area or region of the body, responsible for distributing the vital energy to the organs and parts of that particular region of the body. For such distribution of energy to all parts of the body Chakra System has tens of thousands of thread-like channels called 'Nadis' through which all the centres of energies are interconnected.

The ascending order of the seven major Chakras is also related to the development of the levels of consciousness of human being: it may also be seen as a growth, development, of the mind. The highest one among the seven Chakras, is located on top of the head and is called the 'Crown Chakra' which is the ultimate aim of every Kundalini meditator to attain to.

According to ancient Chinese medicine there are 12 pairs of meridians running deeply in the skin of human body. Along these meridians there are points or inlets, known to us as acupuncture points. Through these points a nutritive substance known as 'Chi energy' is distributed to all parts of the body. Each meridian (pair of meridians) is responsible for distributing the Chi energy to a particular area (parts, organs, etc.) of the body.

Underlying acupuncture meridian system is the philosophy of 'Yin and Yang' two seemingly opposite but in reality two complementary aspects of energy force. Yin is passive and of female characteristics, and Yang is active and of male characteristics. According to this philosophy the concept of Yin-Yang applies to the whole universe and to a wide-range of aspects in life and of course also to human being. The right balance and equilibrium between Yin and Yang in a human being indicates the harmony and well being in that person. The whole aim in acupuncture therapy is to bring about such equilibrium and to re-establish the right balance in the flow of energy along the meridians. This is done by inserting special needles (into the skin) along the particular meridians that have to be dealt with; this is carried out by an experienced Acupuncture practitioner.

I have already said that human body is the 'ground-field' of energy. Energy is flowing, non-stop, throughout the body, nourishing all parts and organs of the body even at cellular level. I only wish the Western medical science would accept this vitally important biological (and anatomical) aspect of human being. In all approaches of healing, including Self-healing, these energies play the most important role.

According to energy medicine most of the illnesses and diseases are largely due to a diminution or fluctuation (or perhaps a blockage) in the normal flow of the energy in the body. In any approach to energy medicine the physician, therapist or practitioner aims at removing such abnormality in the flow of energy in the body. This is done by reactivating the flow, and re-establishing a 'perfect balance', of energy in the person concerned. Though on its own is not holistic medicine, Energy medicine has some holistic qualities; that is why it is central to holistic medicine and essential to holistic treatment.

Although Western conventional medicine is still reluctant to 'come out' and openly acknowledge the existence of human subtle energies, there is a chink of hope across the horizon! Many medical doctors in the past few decades have been taking some notice of human bioenergy and its vital importance in the health and well-being of human beings; they are thinking, how these energies may be used for diagnostic and even perhaps for curative purposes. In fact there have been a number of equipments, energy devices, constructed for such applications in some of the Western countries where these devices are becoming widely used. It is interesting to know that great majority of these equipments are used by medical people.

Today through great technological advances and by designing special equipments called 'electro-photography' these energies can be photographed. I have seen an electro-photograph of the major Chakras taken by a friend of mine who is a well-known researcher and therapist in the field of 'bioenergetics'. We must do more and more study and research into this field, because the greater is our knowledge of human subtle energies and our understanding of their applications for the well-being of human beings, the deeper will be our insight into the nature of the 'Healing Resources' of this multi-dimensional being called human being. Such Science comes under the 'umbrella', what is now called, *Life Field*. And let us hope time will come that we will have a truly 'Healing Medicine' ------- a Multi-Dimensional Medicine, which will also include the technologically advanced (and purely physical) Western Conventional Medicine.

It is interesting that in the philosophies of both energy systems that have been discussed here, we read that these energies are universal ---- *cosmic energies*. Indeed, human subtle energy, among other things, is also a 'connecting link' between human being and the Cosmos. Another such 'link', connecting link, is human consciousness.

9

The Non-biological Aspect of Cancer

There has never been a more frightening disease than cancer, which really is not a single disease; a large number of diseases come under this dreadful 'umbrella' called *cancer*. This essay is not about the biological (and the over-emphasized carcinogenic) aspects of cancer, but chiefly about the non-biological aspects of it. In this essay the main (the main but not the only) aim has been at dealing with the relation, directly or indirectly, between the mind and this thing called 'cancer'. That is to find out the role that some of the psychological factors and certain psychosocial elements mostly of emotional nature play in the development of cancer----- i.e. in the main, the psychosomatic aspects of cancer are dealt with. There are also a few other issues (though may not be quite non-biological) that have been discussed in relation to cancer.

That life fundamentally is hard, and that the modern life-style especially in the West is becoming increasingly stressful is a fact. The more technologically advanced become the Western societies, the greater becomes people's desire to possess more, and thus they strive strenuously to own more and more materialistic possessions. In general Western people care more about wealth than health, though not necessarily in a conscious way; the materialistic living and the competitive way of life dictate to them to behave thus. In fact a large number of people particularly the youth strive to attain even the unattainable. In these societies the stressfully hectic life often seems like an 'aimless struggle'---- struggling for the sake of struggling. It is not therefore surprising that in the past three decades or so there has been a drastic increase in psychosomatic illnesses, nervous diseases, mental illnesses, drug addiction, alcoholism, and also in the number of suicides especially among young people. Furthermore, there is ample evidence based on the studies and researches carried out during past four decades or so that there exist a correlation between certain psychological factors and psychosocial elements and the incidence of cancer. This essay consists of two parts: *Psychological risks of Cancer* and *Life-Style and Cancer.*

Let me, however, try to define the term 'psychosomatics', I should define it thus: the term *psychosomatics* refers to the interrelation and interaction of mind and body, with an emphasis on the mind's influence on the body. I will also try to define the term *psychosomatic disorder.* My definition would be, any illness or disease that certain psychological elements have contributed to it's development or growth is psychosomatic disorder. This means, as we can see, the question whether a disease is psychosomatic or not is all a matter of degree. The greater is the role of psychological elements in giving rise to the growth of an illness the more psychosomatic the illness in question is.

Subtle Energies and Cancer

Some readers may ask, "But what does subtle energy have to do with cancer or with psychosomatics in general?" Well, it has everything to do with it. In fact my 'model' of cancer is a 'bioenergetic' (pschobioenergetic) one. Here I don't want to say much about this as I have written a book on *Cancer* which is going to be published in near future. However, I am going to say only a few words on the subtle energies in human being (human bioenergies) and their close relation to psychosomatics, hence, their relevance to cancer.

In the West psychosomatic researchers, when discussing 'psychosomatics' or writing on 'psychosomatic disorders', they forget or ignore altogether one element that is essential to the whole thing, and that is the bioenergy of the body. Their model of human mind-body interrelation, psychosomatics, is two-dimensional ---- Mind and Body. This, in my view, is not a true model of *psychosomatic phenomenon*. My psychosomatic theory is three-dimensional (Mind Body and Subtle energies of the body). Human bioenergy, the subtle energies, also, among other things, act as an *Interface* between the mind and the body. The mind-body interaction takes place through or via the bioenergies of human being. To talk of psychosomatics and pychosomatic disorders without talking about, or at least mentioning human subtle energies is like when talking about 'driving' pretending that all that matters is the driver and the car, petrol matters not!

Psychological Risks of Cancer

During last few decades there have been constant warnings against carcinogenic aspects of cancer in general and against smoking in particular. Such campaign, warning against smoking, has had its positive effects; there has been some diminution in 'fumy appetite' among the population. The term 'carcinogen' refers to any substance that is capable, when consumed (usually regularly), of giving rise to cancer among the consumers of that particular substance. For example, smoking (the 'nicotine' in tobacco) is said to be carcinogenic, i.e. capable of causing lung cancer. The warnings regarding cancer have all been about the carcinogenic risks of cancer, but (you may well ask) what about the psychological risks of cancer? In truth, the medical system in the West has never wanted (or has always been greatly reluctant) to know about, let alone to accept, the 'psychological risks' of cancer. This is despite of the fact that during past forty or fifty years there have been many studies and researches into psychosomatic aspects of cancer, all supporting the notion (the fact) that psychological factors play an important part in the development of most types and great majority of cases of cancer. There is plenty of literature available for anyone who is interested in knowing about the subject matter in question. Yes Indeed, today it is more or less an accepted fact (accepted not only by the psychosomatic researchers in this area but also by a 'slowly growing number' of medical people) that certain psychological stresses and psychosocial strains in the lives of many people over a period of time can precipitate the onset and also are capable of contributing to the development of many types of the diseases called cancer; in my view this applies mostly to cancers with solid tumour.

Based on ample evidence the research findings have shown that in the lives of a large number of people certain episodes, life experiences over a period of time, are capable of precipitating cancer in them. I am going to deal in brief with a few of the notable stressful life experiences mostly of emotional nature that are capable of precipitating cancer; these often are important contributory factors in the development of cancers mainly with solid tumour.

Helplessness-Hopelessness. According to the studies and researches into the psychosomatic aspects of cancer probably the most common 'emotionally stressful state' among cancer sufferers has been the feeling of *Helplessness-Hopelessness* in their lives; many of them have experienced this in most part of their lives. This is a particular 'despairing state', a 'condition', often referred to as a *sense of giving up*; it is believed, as far as psychological risks of cancer are concerned, to be 'perhaps the most prevalent' contributory factor to the development of cancer. Perhaps most of the researchers concur on this view.

Suppression of emotions. This is probably as common as the previous condition, helplessness-hopelessness. This psychological state, the suppression of emotions, is, to put it crudely, 'the bottling up' of negative and unhealthy emotions and emotional feelings. Some researchers have maintained that this condition, the state of living with suppressed emotional feelings, perhaps is the most damaging when it comes to contributing to the 'onset' of cancer in a person. One who has over a period of some years 'bottled up' in oneself negative emotions and unexpressed emotional feelings (feelings of anger, revenge, jealousy, bitterness, etc) have rendered oneself susceptible to different kinds of illnesses even to serious diseases including cancer. Indeed, this particular psychological (mental and emotional) state probably has been experienced by most of the cancer sufferers in most part of their lives. Some researchers have said that suppression of emotions among cancer sufferers goes back to their childhood; they have always had an inadequate ability in expressing their emotional feelings. In other words, the inability to express the emotional feelings, when young, can easily lead to the suppression of negative emotions in adulthood. It is believed that the most destructive suppressed emotion is 'anger'. I too believe this; bitterness and kept anger probably are the most damaging to both mental and physical health (and in my opinion more damaging to men than to women).

Loss of close emotional relationship. This too has been experienced by a large number of cancer sufferers. Such loss could be the death of a loved one, the end of a close relationship or perhaps the loss of 'something' that meant a great deal to the person concerned. Many people with cancer have experienced more than one loss. And they all talk of such losses with deep sorrow and say how emotionally tormenting such experiences were. Furthermore, from the studies and researches carried out into this area we come to know that this kind of 'emotionally devastating experiences' (the loss of emotionally close relationships) have had their highly adverse effects on their lives in general and on their mental and physical health in particular. I attest this view on the ground of the fact that most of my cancer patients and other cancer sufferers whom I have known or met had had at least one loss of a loved one or of a close relationship (some of them a series of such losses); they all spoke of the devastating effect of these emotional experiences. The studies of Lawrence Leshan on *'Cancer and the Loss of Close Relationship'* are interesting and significant. According to Leshan, the losses of these meaningful (meaningful to cancer sufferers) and loving relationships were greatly traumatic in the lives of his cancer patients, and left terrible emotional scars in them.

Lack of close emotional relationship. The lack of close emotional relationship or the absence of the 'feeling of love' is probably the most painful emotional feeling for human beings. Unfortunately, not many (by comparison) researches have been carried out into this 'sensitive' area. However, The studies show, and commonsense tells us, that the 'absence of love' can be, and I believe often it is, a major influencing factor in rendering people prone to various kinds of ill-health, even serious diseases. In general, the 'absence' of Love in a community or society is in direct proportion to the 'presence' of materialism in that society. Sadly, in many parts of the World, especially in the West, materialism is so prevalent that the constant struggle for the acquisition of more and more materialistic possessions has become a 'norm'. Today there are so many people who see the world as cruel as it can be and perceive life as uncaring life; they feel frustrated and utterly unwanted. In them there is a deep 'sense of emptiness'. Frustration and sense of emptiness are usually the 'best ingredients' for rendering the person concerned, as has just been said, susceptible to illnesses and sometimes to serious diseases. Unless they work on themselves by using suitable approaches to personal growth, self-development, or self-discovery, these people (most of them most probably) will resort to heavy drinking, drugs and also some of them sadly to suicide. Self-discovery, if aimed at diligently and with commitment, is a 'Path' on which the person will, among other things, gradually acquire a realization to his or her *inner self* and progressively will find insight into their *higher, spiritual, dimension*.

Carcinogenic Truth.

As has been said earlier, we are not told about or warned against the psychological risks of cancer, but constantly we are reminded of the dangers of the 'carcinogens' (cancer-producing substances) particularly in relation to smoking. There are things that may be carcinogens, however, the danger of such substances (carcinogenic risk) has been exaggerated and exaggerated in a distorted way. So, what is the truth about such risks?

Only in some people and in a few kinds of cancer a carcinogen, on its own, can produce cancer. In most cases and great majority of types of cancer the so-called carcinogen cannot create cancer; it needs the 'presence' of other agents or elements to produce cancer. And most 'notable' agent that can, and usually does, assist the carcinogen in giving rise to malignant tumour is of psychological, especially of emotional, nature. Here I am going, as briefly as possible, to illustrate the truth about carcinogens and the 'vilified villain' called 'carcinogenic substance'. This will also tell us more about the extent of psychosomatic aspects of cancer. The best illustration of this kind would be the issue of *Smoking and Lung Cancer.*

Smoking and Lung Cancer. The continual warnings against smoking during the past three decades or so have been conducted, on the whole, seriously and at times zealously. And the campaign has had some positive effects; there has been some decrease in the number of the smokers in Britain, however, some may not accept this.

Nobody with a reasonable degree of common sense will dispute that smoking is a 'filthy habit' and also it is unhealthy and even in some cases it can be a danger to the health of the smokers. But what is the truth about the link between smoking and lung cancer? This is what I am going to discuss briefly. Is it an exaggerated link? Could it be a distorted link? Well let us see.

Only in some people smoking can be, and probably is, the 'single cause' of lung cancer. In great majority of smokers, smoking (the substance called 'nicotine') needs the 'presence' of some other elements in order to give rise to the cancer of lung, and the element in question in most, but not in all, cases is a psychological one. According to the findings of the researches carried out into this area, poor emotional outlet in the smokers is the 'main collaborator' with the tobacco nicotine in creating lung cancer in them. Indeed, common sense tells me that in most cases, if not in all, there must be also 'something else' at work (apart from nicotine) in causing lung cancer among the smokers. **I say this because not all smokers get lung cancer; neither are all non-smokers immune from lung cancer.** Today there are tens of thousands of people, if not hundreds of thousands, in London alone who have been smoking for forty years (and perhaps more) at least forty cigarettes a day. Many, perhaps most, of them will go on living a long life and when they die it is not from lung cancer, or any other kind of cancer for that matter. In fact some (perhaps many) of them have hardly suffered from serious illnesses or diseases. Yet, the young man of twenty-four who had smoked for four years only, and not more than fourteen cigarettes a day died from lung cancer, and according to the doctors his death was caused by smoking. But why? Is there any explanation for this? I have never heard or read any medical explanation as to why so many non-heavy smokers die every day from lung cancer, yet there are people who have been heavy smokers throughout their lives and when they die at old age (some of them eighty or perhaps ninety), their death is not from lung cancer or any other kind of cancer. Why? Surely there must be an explanation for this unfair 'cancerous affair'. Actually there is.

According to doctor Kissin, a well-known researcher in this area, there is an explanation for it. Doctor Kissin, who many years ago (1963) conducted researches and did thorough studies of a large number of cancer patients, came up with some interesting findings. He said that the main reason for this is that great majority, if not all, of the smokers who die from lung cancer are people who in their lives (some of them even in their childhood) have not had adequate emotional outlet; i.e. poor emotional outlet has been part of their personalities. Kissin maintains these persons lack sufficient capacity, the adequate ability, to express their inner feelings, thus suppressing them (their emotional feelings) all the time. We are back again to the 'suppression' of emotional feelings which plays, as has already been discussed, an important part in the psychosomatic aspect of cancer.

Furthermore, let us remember that a large number of people who have never smoked a cigarette in their lives get lung cancer and finally die from it. My mother, a healthy woman and a non-smoker, died at a relatively early age from lung cancer. The notion that every lung cancer is caused by inhaling the smoke of cigarette is absurd and nonsense, and even more nonsensical than that is **the silly theory called 'passive smoking'**, which roughly says, "I smoke here, and you get cancer there". In fact the phrase 'passive smoking' is stupid and ridiculous; it implies, more or less, that every lung cancer is necessarily from inhaling the smoke of cigarette. Then what about other kinds of filthy smoke in the polluted air, some of which are by far worse, more dangerous, than the smoke of cigarette? And even to say that every lung cancer is 'necessarily' created by the exposure of the lungs to the smoke (any kind of smoke} is a fallacy. Cancer is highly complex and a multi-causal disease. *Just because of not being able to understand the causes of cancer, and just because of feeling helpless to do something about this 'terrifying thing' called cancer, there is no need for medical system to invent illusory gambits and hide itself behind its delusory theories.*

Smoking is without any doubt unhealthy and capable, over a period of time, of giving rise to respiratory malfunctions and contributing to heart diseases and even in some cases to lung cancer. As a non-smoker I wish to see the day that everyone, all human beings, breathing smokeless, unpolluted and fresh air. Let us hope this is not just a wishful thinking. Having said this, I must say again that the danger of smoking cigarette in relation to lung cancer has been distortedly exaggerated; also, let us, we the non-smokers, stop being so sanctimoniously contemptuous of the smokers.

What about breast cancer? Is it of psychosomatic nature? Yes indeed, at least in most cases, but of a 'peculiar' kind. The psychosomatic aspect of breast cancer is not so straightforward; it is more of an attitudinal nature. A woman's attitude to herself as a woman, and her acceptance of her womanhood especially of her female instincts and her perception of her interpersonal relations largely determine her susceptibility to breast cancer. Considering these influencing factors, it would be reasonable to infer that breast cancer is largely an emotionally related and partly a hormonally- linked disease. This is supported by the findings obtained from the studies and researches done into this area. For example, women who marry at an early (relatively early) age and have a few children (pregnancies) and especially 'enjoy breast-feeding' their babies are highly unlikely to get breast cancer. This is evident from the difference in the rate of breast cancer incidence between Eastern and Western women. Breast cancer by far is less among the Eastern women than among Western women. Western women, generally speaking, when it comes to the age of marriage, the number of children (pregnancies), and particularly breast-feeding their babies, are almost the opposite of the Eastern women. In many parts of the Eastern World breast cancer is rare and even in certain parts is unheard of. In the East women accept they are women and enjoy their femaleness (their being women). But in the West, most women, in many respects, still rebel (not so much openly but dormantly) against their being women (their 'femaleness'). Such 'rebellion' by women against their instinctive nature plays important role in precipitating in them, or at least in rendering them susceptible to, breast cancer, which, in a way and generally speaking, is the 'cry of unsatisfied breasts'.

The Spiritual Dimension and Cancer. A multifactorial disease such as cancer involves different dimensions and includes various levels of human being. So, there is also something, to say about the spiritual dimension of the cancer sufferers. Human beings usually (usually but not necessarily) after having, over a period of time, been through their afflictions, become closer to or aware of their spiritual dimensions. Cancer perhaps more than any other disease brings its sufferers to their spiritual realization. Many, probably most, of cancer sufferers somehow relate the spiritual aspects of their lives to their cancer. They often associate their pain and suffering from cancer somewhat with their spirituality and somehow with their spiritual needs.

A Psychosomatic Conclusion

On Cancer

Now let me summarize my conclusive view in a few words on the Psychological risks, specifically the psychosomatic aspects, of cancer. It is as follows,

> **In great majority of cases in the onset and development of <u>cancer with solid tumour</u> (and also to a large extent leukaemia among adults), the psychological factors and the psychosocial elements chiefly of emotional nature play by far greater part than any physical substances or**

biological agents do. This process, like any other psycho-somatic process, takes place through the subtle energies of the body.
(In my next book, on psychosomatic aspects of cancer, I have explained in reasonable detail the significance of human bioenergy in relation to the growth of cancer)

Life-Style and Cancer

The term 'life-style' simply refers to an individual's way of life ---- the way a person lives. One's life-style consists mainly of one's diet, social life, pattern of sleep, habits, vocation, pursuits and interests, hobbies and leisure, and so on. All these together, collectively, make up one's life-style. In short, 'life-style' is the particular way of life of a person. Often a stressful life-style 'invites' and a disorderly way of life 'attracts' illnesses and sometimes also serious diseases including cancer.

People, in general, learn their life-styles from the grown-ups around them and later (as fully mature adults) also from their friends. However, this does not mean that we are justified to abdicate our responsibilities when it comes to our life-styles. As adult human beings we are responsible for our life-styles. There are of course certain situations in the lives of some people that are beyond their control; even so, it is our duty to endeavour to maintain some control over, or some balance in, our lives. In short, it is the obligation of every person to put his or her house (which includes life-style) in order. This means it is important, even essential, to adopt as far as possible a 'constructive' (sensible and healthy) life-style in order to lessen, even perhaps minimize, our susceptibility to illnesses and many of the serious diseases.

Diet. A healthy diet is very important. A healthy diet is balanced in quantity and in quality and has more or less the nutrients essential to maintaining a healthy body. Such diet is, as far as possible, 'natural' and free from processing, chemicals and additives and it contains only a little salt, not much sugar and a moderate amount of fat. Natural food is a food that is truly healthy and is incapable of causing ill-health or having adverse side-effects. A healthy diet must not be restricted or boring, but it should be appetizing and consisting of a wide range of foods including (if the person is not a vegetarian) a moderate amount of meat. Healthy diet does not mean, as some people believe, 'vegetarianism'. Leaving aside its moral and philosophical implications, the eating of meat in moderate amounts, once or twice a week, is not unhealthy, but on the whole highly nutritious and beneficial. Having said this, I feel 'conscientiously compelled' to say that eating a lot of meat is damaging to the health and even in certain cases, over a period of time, can and probably will render the person susceptible to illnesses and certain diseases. Vegetarians, on the whole, live longer than non-vegetarians. (I am not a vegetarian, nor am I a 'big meat-eater'.) Not only vegetables but also fruits must be part of a healthy diet. As part of cancer prevention programme, apricots, grapes, oranges and cherries are recommended. Some years ago as part of my psychological treatment for cancer patients I recommended the consumption of some fruits and vegetables in the morning as part of (or preferably before) breakfast. And I recommend the eating of raw (as raw as possible) vegetables---perhaps everyday.

Sleep. One must sleep sufficiently. A normal sleep pattern is very important even essential to the health of a person. Insufficient or the irregular pattern of sleep is unhealthy and over a period of time can adversely affect, in some individuals seriously, the immune system of the body. Any deficiency in the immune system gradually, over a period of time, will result in a major dysfunction in the performance of the defence mechanism of the body, thus rendering it unable to fight the adverse effects of the harmful elements and to eliminate the abnormal cells in the body. To put it simply, a weak or mal-functioning immune system cannot defend the body against the 'harmful invaders' and cannot fight effectively the cells that behave abnormally. Now we can see the importance of *Good Sleep*. And let us remember that insufficient sleep or the irregular pattern of sleep is not the only thing that adversely affects the immune system of the body; there are a number of elements that are capable of doing this and some of these are by far worse. Without an efficient defence mechanism the body becomes 'exposed' to all kinds of problems related to health and well-being. Such problems can, and often do, lead to illnesses and at times to serious diseases. In maintaining health and sustaining a 'sense of well-being' good sleep is important and a regular sleep pattern matters a lot.

Alcohol. This is a strong drug, which, if consumed in excess, most probably, over a period of time will begin to deteriorate the health and will adversely affect the state of well-being of the person concerned, the drinker. In some cases, in some individuals, such deterioration of health may lead to serious health problems, even complicated diseases. Being hooked on alcohol in majority of cases can be, and very often is, devastating; indeed, alcohol, more than any other drug, has ruined the lives of people; this is prevalent especially in the Western world.

Most 'heavy drinkers' (a euphemism for "respectable" alcoholics) began drinking excessively as a way of 'escaping' from their situational problems. Indeed, most of them resorted to such 'escapism' in order to sooth, perhaps to 'anaesthetize', their unhealed inner wounds and emotional scars. In life, however, there are much better ways of dealing with the stresses and strains of life, and with the 'unhealed wounds' of the inner psyche. We ought to learn and apply them in our daily lives.

Human Suffering. Buddha started his teaching by saying that life in this world is full of disappointment and suffering. In fact in Buddhism, according to Buddha, suffering, mainly disappointments, frustrations and dissatisfactions are inherent in life. Jesus in a way was perhaps more emphatic and explicit when it comes to our pains and sufferings. He says (in St John's Gospel) "You will have many troubles in this world." What about His repeated statement----"Take up your cross and follow me". The word 'cross' in such contexts denotes human 'pains and sufferings' in this world. Though both in their teachings concur on the existence of human suffering, Christ and Buddha have very different views in dealing with suffering. (I am not going to discuss this here; this is not the place for it). No human being wants to suffer. However, life is hard, and human beings do experience pain and go through their suffering, though in different ways and in varying degrees. When faced with difficulties and problems, we must be prepared to confront them courageously and go through the painful experiences acceptingly. I say this because in every person's afflictions there are 'hidden truths' about that person. Indeed our sufferings 'contain' many truths about us. To run away from our sufferings, is to run away from the realities of our lives, and to turn a blind eye to our pains is to turn a blind eye to many truths related to our 'being' in this world. But by enduring our sufferings and by persevering when going through painful experiences we will begin to know about ourselves and will certainly acquire an understanding of our 'Self' that hitherto we knew not. It is a process of Self-development, which may lead progressively to 'Self-realization'. To reach such 'stages' in life is somehow fulfilling since it enables the person, among other things, to find some meaning in life.

Purposeful Life. Life must be lived with and for a purpose, a good and positive purpose. A purposeless life is an 'empty life' and especially to live aimlessly is a frustrating life. Such a life can and usually does render the person susceptible to almost every kind of ill-health, not only physical but also psychological ill-health. Indeed, to live with 'special aims' and a positive purpose' not only is healthy, but also it often helps in preventing us from being affected by illnesses; moreover, it is fulfilling as well as rewarding.

Interests and Hobbies. Let me also say a few words about 'interests and pursuits'. Enjoyable pursuits and healthy hobbies will enhance the quality of life, and pleasant interests will enrich our life-styles. Such pursuits and interests don't have to be expensive and spectacular, but ordinary and simple, because usually it is the simple and ordinary things in life that cherish the Soul most.

<u>Finally</u>, let us feel the sorrows as well as the joys of life. Life must be felt and felt in all circumstances especially in its intensities. They lived who felt acceptingly both the sadness and the gladness of life. Indeed, this is the way to grow, this is the way to develop, and this is the way to learn about the 'growth of the Soul' and to acquire wisdom. To experience the pain (yes, even the pain of cancer) and to cherish the joy of success are both part and parcel of life; this is an indisputable fact of life that ought to be accepted.

10

Mind-Brain Problem:
A wHolistic Dualist Theory

One of the most perplexing problems in philosophy, which has been with us for thousands of years, is the Mind-Body ('Mind-Brain') problem. Even to this day the protagonists of the extremes argue among themselves so vehemently - each side trying to convince us that they are right. One group says, there is no such thing as mind; it is all brain - mind is nothing more than a delusory phenomenon. The other group says, it is only mind; matter is an illusory thing. And of course between these two polarities there are many views, concepts and theories of 'mind and brain issue'. Writers on the subject in question like to bring all these views and theories under two 'umbrellas'--- 'monism' and 'dualism'. I will not do so; only briefly I shall deal with some of them and then I will put forward my own theory of mind-brain problem.

Let us first ask ourselves how much progress have we, the researchers in such areas, made in solving the problem of mind and brain. The honest answer is, 'None'. Human being is still as remote from knowing the truth about the working of the mind and brain as they (human beings) were centuries ago. Some years ago in a book on human mind and related areas I read that we still do not know how the thought is 'converted' into a 'mechanical' (physical) activity. For example, I just 'thought' of raising my left hand. How am I doing that - because I honestly don't know. Now I decided to stand up. A second later I stand up. Who can explain the 'psycho-neurophysiology' (as a process) that went on (took place) in me from the time I 'decided' to stand up to the time that I was in 'standing-up' position? No wonder one of the most eminent neuroscientists of the 20th century in a conference, addressing a large audience mostly scientists, said that the greatest thing the scientists can achieve will be to explain the interrelation, 'the working interrelation', of mind and brain. And the same neuroscientist once said to the audience (perhaps in the same conference?), while moving his index finger, 'How am I doing this?' - ---- suggesting that we are unable to explain in terms of mind-brain process even the moving of our index finger.

I accept that the question of 'Mind-Brain issue' is a tough one, a very tough one indeed. In fact we may not be able to have even some reasonable understanding of the problem in question, let alone to find a way of solving it. Having said that, let me also say that our lack of making a good progress in this field has been partly because of the over-zealousness of the 'purely mentalists' who overlook matter (as if it did not exist) when arguing their case, and also largely is due to the hard-headedness of hard materialists who can not stand the word 'mind' and have no time for the word 'consciousness', and the word 'self' is alien to them.

Intoxicated in the ocean of their obstinacy, staunch materialists believe that nothing can exist apart from that which can be seen, or touched, or in short, if it does not have a physical reality. To these people existential reality is only in physicality. They are desperately clinging to their 'crumbling matter' despite the fact that the great man, Einstein, showed by his famous equation, $E=mc^2$, that matter is the 'storage of energy', and even mass and energy are 'convertible'. And what about Quantum physics, in particular its 'wave-particle duality' which debunked the 'reality' of any solid building block for matter.

Before dealing with some of the theories and concepts of 'mind and brain issue', let me sweep away a popular misconception: it is not only believers in God, or spiritual people who believe or accept the reality, the existence, of a non-physical entity such as consciousness or human mind. The late Karl Popper, the distinguished philosopher of science and a self-confessed agnostic, believed in human Self/Soul. Moreover, Popper believed that brain belongs to the Self and not the other way round.

Some of the Theories on Mind-Brain Problem

Mind-Brain Dualism: This is one of the two main categories used in the classification of mind-brain theories. There are a number of dualist theories. A dualist theory holds the view that mind and brain are two distinct substances, or properties. Though it would oppose any theory that identifies mind with the brain, a dualist theory is not necessarily a non-materialist theory. Epiphenomenalism, for example, though comes under 'dualist theory', it is very much a materialist theory.

Mind-Brain Monism: Any mind-brain theory that denies the duality of mind and brain (or believes the two are a 'single property') is a monistic theory. Two very famous or 'highly notorious' monistic theories (each at one end of the monistic spectrum) come to my mind: one says, it is all mind and it denies the reality of matter, and the other one says it is all brain and it denies the existence of mind. The former is associated chiefly with 'Berkeleyan theory' and the latter is famously known as 'Identity theory'. I am going to describe them very briefly.

Berkeleyan theory: The theory is named after Bishop George Berkeley. It is a purely mentalistic concept of mind and brain. Berkeley insisted that it is all mind. He virtually denied the reality of matter, saying, roughly, physical objects are the collection of sensible qualities, and they can be explained, he said, in terms of mind. Many Berkeleyans say, "No, it is not as simple as that; there is much more to it than that'.

Identity theory (Reductive Materialist theory): This theory is materialist in its extreme. It says all mental events are nothing but physical events (neuronal events); and every mental state, process or experience, is capable of being explained in physical (neuronal) terms. In short, mental states are brain states.

Epiphenomenalist theory: This theory says that mental events are the by-product of brain activities, neuronal events; however, these mental events do not have any causal efficacy on the brain whatsoever. In other words, according to this theory, it is all 'one-way traffic'. T.H. Huxley was the well-known proponent of Epiphenomenalism.

Interactionist Dualism: It is a dualist theory which holds the view that mind and brain are constantly in the state of interaction. John Eccles, the well-known neuroscientist and a Nobel Prize winner, was a strong interactionist. Eccles and Popper (the philosopher), together wrote a book, 'Self and its Brain' - probably the best book written on mind-brain interactionism.

Parallelism: This is the theory which says that mind and brain function independently in parallel - mental states and physical states run in parallel; therefore, they cannot interact.

Emergentism: This theory was introduced by **Roger Sperry**, the well-known American neuroscientist and a pioneer in hemispheroctomy. Sperry says that it is the brain, specifically the dominant hemisphere, that gives rise to consciousness, and this consciousness has great control over the brain. Put another way, in Sperry's view (and he bases his views on his professional experience including his experiments in this area), mental events emerge from neuronal events, but the mental events themselves have causal influence on neuronal events.

Neutral Monism: This theory was advanced by **William James** and then by others notably **Bertrand Russell**. According to this theory, it is 'all one thing'. It says that mental events and physical events are two sides of the same overall process; there isn't a real division between them. It is said that Neutral Monism emerged from Berkeleyan Mentalism and later it 'rubbed shoulders' with Hume's Bundle Theory.

Physics and Consciousness.

Physics is the science of matter, physical things. A physicist is a physical, material, scientist. He or she tries to know more and understand better about things by studying and researching into material objects. Ambitious physicists endeavour to discover truths and uncover realities by diligently investigating into matter. It was, however, by the turn of the century (a hundred years ago) that something began, gradually, to change the science of physics. A 'realization' began to dawn on the consciousness of some of the physicists in the West. They began to realize, not necessarily in a conscious manner, that the truth and the reality are not the monopoly of the matter, nor the investigation into matter is the only way of finding the truth. These scientists progressively became more and more confident that *Reality* is (also) beyond matter. With that realization and with such confidence they went on doing their works. **Therefore, Quantum Physics, as we know it today, has**

been the product of the Progressive Realization of the Minds, the Consciousness, of those great physicists, the Quantum Pioneers, followed by the quantum-related concepts and theories of a dozen or so top physicists of the second half of the 20th century. .

That is why the changes or developments especially in the past fifty or sixty years that have taken place in the area of human consciousness have mostly been the product of the diligent research of a number of physicists, chiefly quantum physicists. However, these developments are also in part due to the studious endeavour of many scientists from other fields; these are mainly neuroscientists and psychologists. Probably the most significant development taken place in this area in the recent decades is the theory or the concept that says 'consciousness' is a 'non-physical' entity ----- both non-physical and non-local; and as a "hidden variable" it is working in conjunction with the brain.

As for the non-physicality and non-locality of Consciousness various theories have been put forward mainly by a number of Quantum physicists. One such physicist that comes to my mind is the late **Evan Harris Walker** whose early researches in the area in question go back many years. Walker was a scientist of eminence who worked for at least thirty years at different research projects including the NASA Electronics, USA. He did extensive research into consciousness and its relation with brain (mind-brain issue). His work in this field is scientifically comprehensive, and it involves certain Quantum mechanical equations and some other elaborate mathematical works. His contributions to several scientific fields have been significant. He wrote perhaps more than a hundred papers, all published in scientific journals, and he also wrote a number of books; his last book (I believe it is the last) is *The Physics of Consciousness: The Quantum Mind and The Meaning of Life.* Reviewers of his books and critics of his scientific works differ in their views and vary in their criticisms. To interpret Walker's concepts and to assess his conclusions, some of which are paradoxical, is not an easy task. Here, in a paragraph, I am going to outline briefly some of his views and concepts on consciousness and brain in terms of Quantum Mechanics.

Evan Harris Walker, like Einstein, knew Quantum Mechanics is incomplete, filled with probabilities and uncertainties. However, unlike Einstein, he did not keep away from it; in fact he became a true Quantum theorist, because he saw the 'good things' of it, things that gradually would enable the physicists to develop a complete 'New Physics'. As a physicist, Walker loved to see a New Physics that would embrace Consciousness as a *Reality*. To him such reality, the reality of consciousness, is non-physical. And of course he is not the only one who sees consciousness as such. Walker says, **"Consciousness can not be found among physical objects"** By this, I believe, Walker, as a physicist, refers particularly to the fact that consciousness cannot be understood or explained within Newtonian physics which is purely mechanistic, whereas the physics called 'quantum mechanics' is not so; there is some room in it to manoeuvre and to search for 'hidden variables' ------ non-physical elements, or non-local realities. Now we can see what I meant, a moment ago, by Walker never stayed away from Quantum Mechanics because he saw 'some good things' in it.

Walker says Consciousness is non-physical reality and is coupled to the brain by way of quantum mechanical wave-function. To explain it better, he says this occurs in the synaptic cleft in the brain, thus producing conscious perception. He maintains 'Electron Tunnelling' across the synapses is the basis for the transmission of the impulses across the synapses. As I understand it, he sees consciousness as the (and I use his phrase) 'quantum-mechanical tunnelling' of the electrons across the synapses, i.e. at the minuscule intersynaptic cleft; **it is there, Walker believes, that the Link between Mind and Brain exists..**

(Evan Harris Walker was born in 1936 and he died in 2006. He was a good man. He lived almost all his adult life with a deep emotional scar in him. He was, among other things, the director of Walker's Cancer Research Institute ------I believe he founded it.) .

A 'wHolistic Dualist Theory' of Mind-Brain Problem (My Theory)

As has been said, Mind-Brain (Mind-Body) problem has baffled the philosophers since ancient times; probably it is the most puzzling philosophical and psychological problem. However, it has throughout been the philosophers who have shown an interest in understanding (and also perhaps in 'solving') this, seemingly, insolvable problem. Psychologists have not considered the mind-brain issue to be an area for psychological research; there have been, however, some, not many, of them who have shown some interest in this puzzlingly interesting area --- how do Mind and Brain function in relation to each other? What is the nature of their working interrelation?

When it comes to mind-brain issue I am an unashamed interactionist, hence, a dualist (but a dualist of a different kind). An interactionist is a dualist, but a dualist is not necessarily an interactionist. In fact most of the dualist theories of 'mind and brain' are not interactionist. I am an 'Interactionist-dualist' of a special (different) kind, and moreover the very title of my theory is 'wHolistic Dualist', which sounds a contradiction in terms, but it is not; I will come to this in a moment.

Though its origin goes back to ancient times, dualist theory of mind and brain is usually associated with Cartesian Concept of mind and body. Cartesian Concept is the philosophic view of the great French philosopher, Rene Descartes (17th C.), known as the father of modern philosophy. And I hasten to say that I am not a Cartesian - perhaps in most respects the opposite.

Descartes said that mind and body are two separate things and they function independently. I say, though two different elements, mind and body (brain) work as a 'unit' and they are completely interdependent and in constant interaction. Descartes said that human body is a physical object obeying physical laws. I say human body is a 'living thing' - a ground-field of energies; moreover, human body does not obey the physical laws, but some psychosomatic (psycho-physiologic) 'rules', and in particular, functionally, it follows certain psycho-bioenergetic patterns. However, let me say (and try to be a bit fair to the great French philosopher) that Descartes later (some say 'too late'), conceptually, 'relaxed' a couple of principles of his mind-brain theory and said that there is some kind of interaction between them. (But I wonder how can there be interaction between mind and brain, if they function, as he said, 'separately and independently'?)

So far I have 'disclosed' my general view on the problem of mind and brain and have 'revealed' my stance, where I stand, in relation to it. Now I am going to describe my 'theory of mind and brain' and will explain it in some detail.

Mind and brain, in my view, are two 'elements' of entirely different (in fact, the opposite) nature: one is material (physical) and one immaterial (non-physical). Though entirely different in substance, mind and brain functionally are completely interdependent. Mind, in order to do its job, needs a functioning brain, and the brain, in order to function, needs the 'presence' of the mind - a 'complete interdependency'. These two entirely different substances (a duality) work as a 'unit' - 'one whole', and function in a wholistic manner, in a 'Holistic way' ----- hence, *wHolistic dualist theory of mind and brain*. This may seem a contradiction in terms, however, it is not a contradiction but a paradox. I will talk of the working of the mind and brain, according to this theory, in a moment.

I maintain that human beings are not born each with his or her individual consciousness. We all (and also the animal kingdom) share the Cosmic Consciousness. The 'consciousness' of each person is the interrelation of the Cosmic Consciousness and the brain of that person. Brain is not the generator, but the receiver (or a sort of 'detecter') of the consciousness. Therefore, though a person does not have an 'individual consciousness' (not born with his or her individual consciousness), he or she has his or her own consciousness (which is the 'product' of the *INTERPLAY* of the person's brain and Cosmic Consciousness). One word of caution! This consciousness must not be confused with the 'Self'. Self or Soul is something that a person is born with; it is this 'special feature' (higher/extra dimension) that is fundamental in differentiating human beings from the animals. The ' I ' grows out of (emerges from) ' Self ' (Soul). It is this extra, higher, dimension (exclusive to human being) that renders human Mind to be conscious of itself; hence, Self-Conscious Mind. Indeed the Mind of human being is conscious of all its (Mind's) activities such as thinking, remembering, planning, deciding, feeling, knowing and so on. Self-consciousness is absent in the animal kingdom. Animals, for example, are aware of their surroundings, but human beings are aware of their surroundings plus the fact that they are aware of such awareness. A rabbit is frightened, but I am frightened and also I know that I am frightened. The cat sees the people in the room; I see the people in the room and also I am conscious of my seeing the people in the room. It is such Self-Consciousness of the mind that enables human beings to think deeply and to think in abstract terms, to innovate, to design, to create and to create masterpieces, and perhaps above all to rise up and surmount their miseries and often to be able to do something about some of the terrible conditions that they and their world are in. Now we know what is the first and foremost thing that differentiates humans from animals (and that is *Self-consciousness*, which stems from the *Self*); intelligence, logic, rationality etc. come later (these are 'secondary' as far as the difference between human being and animal is concerned). Self is exclusive to human beings. Animals don't have it.

As I said earlier, Mind and Brain, according to my wholistic dualist theory, work interdependently, as a unit (one whole). Paradoxically, despite such interdependence and wholeness, Mind and Brain can, 'freely' (yes, freely) exert influence upon each other ---- but with a difference! The difference being that the Mind has by far a greater influence on Brain than brain has on mind. Brain's influence on mind is both conditional and limited. Mind's influence upon brain is not. Mind has a continual and almost unconditional influence upon the brain. Apart from its constant influence and usual effects upon brain, Mind is capable of having, and sometimes has, 'considerable impact' on the working of the brain. In fact, according to this theory, mind's continuous influence and its continual effects on the brain are so considerable that we can justifiably say, as it were, 'the whole show' is run by Mind. Indeed, Mind is the 'primary'. And it is of utmost importance for us, human beings, to believe in and accept the 'Primacy' of the Mind'. I maintain (and I know this from my personal and professional experiences) that the 'acceptance' of the *primacy of mind over brain* (mind over body) can play an important part in our *Self-belief*; this means it will encourage us in believing in ourselves, and in having a greater confidence in ourselves. Furthermore, by acknowledging 'the Primacy of the Mind' we also acknowledge that we are not the victims of illnesses, and we refuse to believe that we are determined by our genes as some overzealous geneticists constantly try to make us believe. Too much credit has been given to the genes as far as their influence is concerned. The working of the genes and genetic influence as a whole is largely dependent on the environmental conditions. Of course there are illnesses or abnormalities that are genetic or at least partly genetic; no one denies this. But please let us stop indulging in exaggerations regarding genetic influences on our bodies and in our lives.

Mind, as has been said, is very much in charge in this theory; but a dependent in charge. Mind is independent, yet a dependent independent. Mind depends on a functioning brain. at the same time there cannot be a functioning brain without the mind. Though I have called it 'wHolistic dualist theory' this theory is not *dualist* in its true sense. I am not a 'confrontationalist dualist' but a *Complementary Dualist*. That is why I am so interested in Chinese concept of 'Yin and Yang', which is a concept of 'amicable dualism'. The working relation of yin and yang is completely complementary. This theory, my *wHolistic dualist theory,* may be somewhat a paradoxical theory but it is never a contradictory theory.

Quantum physicists in general see a close similarity between the Quantum World and the World of Mind and Brain. I am sure my wHolistic dualist theory of mind and brain (with all its interdependencies, paradoxes and also probably 'probabilities') is not a stranger to the Quantum world. And who knows, one day this wHolistic theory of Mind and Brain may be accepted by many people and even it may become the most popular one. Wishful Thinking? Well, let us see.

11

The Story of
Quantum Mechanics

Once a friend of mine who is a scientist and an accomplished physicist, during our conversation about science and related areas, said to me that the simplest way of describing the difference between Newtonian (classical) physics and Quantum physics is that Newton's physics is the physics of large, huge, things and Quantum physics is the physics of small, tiny, things. Though he said it with a smile on his face, while he was having his second (and the last) pint of 'real Ale', he was right, more or less right. Newtonian physics, in the main, is concerned with the motion of the bodies at a large scale and the forces that cause such motions. On the other hand, Quantum mechanics is concerned mainly with the behaviour of the atomic and subatomic elements. Classical Physics, is still taught in schools, and Quantum

Mechanics, which is still enjoying its honeymoon, is enthusiastically lectured and taught in universities. Newtonian physics has throughout suffered from its "insufficiencies", and Quantum Physics is still fighting against its "incompleteness", especially its 'indeterminacies'. Newtonian physics is mechanistic and lifeless, Quantum physics is 'vibrational' ----- filled with energy. Finally, many concepts and ideas of Quantum mechanics are somehow philosophical, and even some of them may be regarded as somewhat metaphysical. That is why it would be reasonable to say that Quantum physics has, at least to some degree, narrowed the 'wide gap' (which has always existed) between Science and Spirit.

Not long ago in a scientific journal I read an article by a German quantum physicist talking about science in general and quantum mechanics in particular. Somewhere in the article he says: '... *and quantum mechanics which was discovered by Planck and Einstein ...* '. 'Discovered?' I said to myself, 'and how can he say it so bluntly that these two men were the discoverers of Quantum Mechanics? I say this because those who know anything about quantum physics will agree that it was not discovered but it was gradually developed. Such development was brought about mainly in the first three decades of the 20th century by a number of eminent physicists, usually known as 'Quantum Pioneers', who were enthusiastically busy doing their Quantum discoveries'.

Still engaged in my internal dialogue I said to myself: the man, in a way, is right, because many aspects of Quantum mechanics were discovered, though not at once. In fact the whole 'development' has been *a progressive discovery*.

And what about Einstein? Discoverer of quantum mechanics? After all he was not, nor is he regarded as, one of the quantum pioneers. Moreover, Einstein detested the word 'quantum' let alone the phrase 'quantum mechanics'. According to his biographer, *'Quantum was his demon'*. Now we can see why Einstein and Niels

Bohr could never get along with each other. The latter could not stand the 'confident and composed greatness' of the former and the former disliked the 'silly-obsession-with-quantum' of the latter. But if Einstein disliked quantum physics and kept himself distant from it, why is it that a number of physicists, some of them quantum physicists, even to this day, say that without Einstein and his scientific contributions we possibly, if not probably, could not have quantum physics. Not only was Einstein a *'scientific inspiration'* for most of the quantum pioneers, but also many of them learned from him and his scientific works. It wasn't only that 'Quantum Doyen', Niels Bohr, who would frequently visit the great man, Einstein, in order to learn something from him, but every one of the well-known quantum pioneers would occasionally consult the 'Lion of Physics' (it is the title that I have given the great man) in order to gain from his scientific knowledge.

It is said that Einstein's 'disinterest' in Quantum physics was partly (or perhaps largely) due to the fact that it involves a lot of probabilities, indeterminacies and uncertainties. Hence, the oft-repeated Einstein's statement: 'God does not play dice'. In fact, this is not quite (according to some books including one of his biographies) what the great man said. During their discussion, Einstein and Max Born were trying to come to some sort of agreement on some of the Quantum issues. Born was endeavouring to explain the probabilities etc. that involved in the Quantum world, to which the great man responded, "You believe in a God who plays dice........ I don't ..." This was what, the first time, he said, but perhaps later on he said it (and I believe he did) to his friends and colleagues in different ways including "God does not play dice'. And he really meant it -------- that God does not play around with the running of the Universe. Though not a religious man as such, Einstein was a "mystically spiritual" man; he was a man always God in his mind. Furthermore, his God was a 'Cosmic God', the God of Law and Order and God of Certainties, not God of disorder and uncertainties. Therefore we can see why Einstein could not accept Quantum Mechanics, and why as far as possible he stayed away from it. This was because the Quantum World is the world of probabilities and uncertinties, and he could not have this as far as his concept of God and his conception of the Universe was concerned. In short, the Great Man could not stand the incompleteness of Quantum Mechanics particularly in relation to its "indeterminate concept" of the Cosmos and also the indeterminacies of the Quantum World itself

It was Planck who laid the foundation for Quantum Theory, and it was Einstein who initiated, paved the way for, the development of Quantum Mechanics, but as an 'inadvertent initiator'. He inadvertently triggered the 'Heart of Quantum Mechanics', which is the 'wave-particle duality'. (I will come to this later). Without Wave-Particle duality how can quantum physics be called 'quantum'? Furthermore, and above all, Einstein was the first person to use the term 'Quanta' (the plural of 'quantum') in a scientific context (in a scientific paper). Planck only passingly made a reference (it was a 'passing reference') to the word 'quantum' in one of his papers. The story is told by OED succinctly.

The greatest authority on etymology and the historical development of words and their meanings in English language is the 20 volumes **Oxford English Dictionary**. In volume XII on pages 980-981 under the entry 'quantum' and under the meaning no.5, we read the account, (a 'triangular account') of **"Quantum - Planck - Einstein"**.

The Story of Wave-Particle Duality

In Quantum Physics 'Wave-Particle duality' is central to quantum mechanics. If we remove the *wave-particle duality* from quantum physics it can hardly (in fact it can not) be called quantum mechanics. In fact *"Uncertainty Principle"*, which is considered to be an important 'Pillar' of Quantum mechanics, was deduced from wave-particle duality. I believe (I speak from my research in this area, thus, it is my subjective view) that not less than fifty percent of literature on quantum mechanics, either directly or indirectly, in one way or another, is related to or associated with the idea or subject matter of wave-particle duality. But how was this phenomenon, wave-particle duality, discovered?

It was in 1905 that Einstein made his highly significant scientific 'announcement', perhaps, not less important than Planck's 'discrete packets of energy', in 1900. Einstein said that light (which was until then thought of as wave only) was not all waves, but also particles and it could be explained in terms of particles. This is related to, what is known as, 'Photo-electric Effect' for which Einstein received a Nobel Prize for physics. *Einstein's discovery that light could be described in terms of 'particle-like quantities', which years later were named 'Photons', became the most important step towards the development of wave-particle duality* - Einstein did not know this at the time.

It is noteworthy (and also perhaps important) to know that Einstein for the discovery of *photo-electric effect* drew upon Planck's 'Quantum Principle' central to which is his equation, E=hf. Einstein used Planck's Constant ('h') in formulating his discovery on the nature of light.

One of the first (probably 'the first') physicists who fully accepted Einstein's concept of light - the photons - was a well-known quantum pioneer, Louis de Broglie, a French physicist. Drawing upon this, de Broglie as part of his Ph.D. research at Sorbonne, Paris, investigated thoroughly the quantum entities and the behaviour of sub-atomic particles. He came to the realization, and later became his major scientific conclusion, that Quantum entities such as electrons can also become waves and then these waves can become particles, and so on. Thus, Louis de Broglie was the first person, among the quantum pioneers, who put forward the idea (or discovered the phenomenon) of 'wave-particle duality' (of course by drawing upon Einstein's

Photo-electric Effect). His research findings could not impress his PhD. supervisor. Therefore, he (the supervisor) showed them to the great man himself, Einstein. Einstein read them, and he was pleased with, and somewhat impressed by, de Broglie's findings. Therefore he 'okayed' the research in question, and this also pleased the fastidious supervisor. Therefore, Louis de Broglie deservedly received his PhD in 1924. Five years later, in 1929, he also received the Nobel Prize for physics. This was of course for his extensive research into subatomic aspects of quantum entities.

Davisson, an American physicist, and his colleague had heard of de Broglie's research findings. They had done a lot of study and research in the area in question. It was in 1927 that they were able, through their scientific experiments, to confirm the validity of the wave-particle nature of quantum entities such as electrons. Such achievement was not without reward. Davisson and his colleague were awarded the Nobel Prize for physics after ten years or so (I believe in 1937}

Quantum Models and Interpretations of Quantum World

There are several models in quantum physics and many Interpretations of Quantum World. None of them is 'the true one'; yet every one of them, most probably, is 'the true one' in a particular 'quantum context'. It all depends to what area or to which concept we apply a particular model. Niels Bohr's model of atomic structure has been accepted 'most widely' probably because it is considered applicable in a wide range of experiments or instances in quantum physics. Bohr was the first physicist to introduce, to formulate, a Quantum Model of atomic structure .

In classical physics the atomic structure was by far simpler, more straightforward, and 'things' were working in a more or less determined and clear-cut fashion. Furthermore, there were not many subatomic particles; now the number of such particles is exceedingly high. In classical physics an atom was thought of as a hard (solid) nucleus, which consisted of protons and neutrons with a number of electrons orbiting in a well-defined manner around it. In Quantum physics the nucleus of the atom consists of a large number of subatomic particles, and the electrons revolve around the nucleus in an erratic way; sometimes they look like well-behaved particles and at times like ill-behaved waves. These 'quantumly eccentric' electrons also jump from one blur orbit to another one seemingly without passing through any intervening space.

Niels Bohr's *Model of Quantum Interpretation*, which later became known as *'Copenhagen Interpretation'*, is now the standard one; it is included in the text-books of quantum physics and is taught as being the best (or perhaps the true) interpretation of quantum world. We don't know of course whether it is the true one or not, however, it seems the model in question to be interesting, intriguing and somewhat, interpretationally, 'mysterious'. Copenhagen Interpretation was constructed by Bohr and a couple of his colleagues together, and was named after Bohr's home and work town, 'Copenhagen'.

Central to this Interpretation is the 'interaction' between the observer and the Quantum world. It says, the observer's conscious observation plays a crucial role in the whole thing. The observer will see what he (or she) expects or wants to see. Put another way (and some like to understand it this way), it is the observer's observation that makes 'the thing' to appear as it appears. It seems Copenhagen Interpretation suggests that 'what cannot be observed does not exist'. There are many who are reluctant to accept this. They find hard to accept an Interpretation that is based on the following Principle :

> **"What is observed certainly
> exists; about what is not
> observed we are still free
> to make suitable assumptions.**

We can use that freedom to avoid paradoxes."

As we can see, it is the observer, or rather the 'observer's conscious observation' that plays the crucial role in the whole thing; and this involves measurement. Bohr's interpretation suggests that reality may only be ascribed to a measurement. Einstein disagrees, suggesting that the physical world has real properties, whether measured or not. Such 'status' given to human conscious observation in the quantum world has intrigued the minds of many scientists particularly certain physicists. One such scientist is the well-known quantum physicist, John Wheeler, who has suggested that the word 'observation' should be replaced by 'participation', and the word 'observer' by 'participator'. Our observation of the universe, according to John Wheeler, is an 'active participation' which plays a crucial role (and I hope I am not mistaken) in the existence of the universe ------- quantumly hot stuff indeed!

As has been said earlier different models and various interpretations have been put forward for the explanation or the interpretation of the 'Quantum Realm'. Among these there is a group of theories referred to as 'Hidden Variables'. One thing that these theories have in common is that they are 'non-local'. The well-known Transactional Interpretation is one such theory, which was formulated by John Cramer, an American physicist. There are many physicists who prefer Cramer's Interpretation to the Copenhagen one.

The late David Bohm, an eminent theoretical physicist, expounded his Hidden Variables theory. Most physicists rejected it then. They did so, according to some quantum physicists, because it made sense more than any other Interpretation! Bohm postulated his quantum theory in 1950's.

Once I read an article on Bohm and some of his works by David Pratt; I recommend it highly. I am going to quote a few lines from it.

> **"Bohm sent copies of his text-book to Bohr and Einstein. Bohr did not respond, but Einstein phoned him to say that he wanted to discuss it with him. In the first of what was to turn into a six-month series of spirited conversations. Einstein enthusiastically told him that he had never seen quantum theory presented so clearly and admitted that he was just as dissatisfied with the orthodox approach as Bohm was. They both admired quantum theory's ability to predict the phenomena, but could not accept that it was complete and that it was impossible to arrive at any clearer understanding of what was going on in the quantum realm."**

In my opinion Bohm was the most underestimated physicist of the 20th century. He was not an eminent physicist only, but also he was a man of high intellect ----- a philosophic thinker. Not only did Einstein admire Bohm for his intellectual capacity, but also it is said that Bohm was the 'favourite physicist' of the great man---Einstein. In recent years, however, there has been a gradual realization of Bohm's 'stature' and of the importance of his combined scientific and philosophic works. He published a few books and many papers. His book *'Wholeness and the Implicate Order'* is unique in its kind - an intriguing paradigm of the universe; it is a holistic concept of the cosmos.

A Brief Chronological List of Discoveries (and developments) in Quantum Physics

In 1900 Planck announced that the radiation from the 'blackbody', emission or absorption, is a discontinuous process; it takes place in discrete packets of energy (i.e. in quanta, which is the plural of 'quantum'). **Thus, Planck laid the foundation (usually referred to as "Quantum Principle") of Quantum theory.**

In 1905 Einstein presented his Special Theory of Relativity Einstein's new concept of light: He suggested that light, contrary to the common belief, is not wave only; it can also be particles, and may be explained in terms of particles too. Many years later these particles were called 'Photons'. This was the beginning of what later became known as the concept of Wave-Particle-Duality which is central to Quantum Mechanics. **In fact Einstein played a most important role, though inadvertently, in paving the way to the development of Quantum Mechanics.**

In 1910 J.J. Thomson, who had already discovered electrons, identified the protons.

In 1913 Niels Bohr constructed his model of atomic structure.

In 1915 Einstein presented his General Theory of Relativity.

In 1916 Einstein puts forward the idea of momentum related to the quantum light.

In 1919 during a solar eclipse measurement was taken of the bending of starlight caused by the Sun's gravity; the result was that it corresponded with Einstein's prediction.

In 1920 Arthur Compton discovers 'Compton effect' --- associated with the increase in the wavelength of certain rays; this is said to confirm the existence of Photons (Einstein's particles of light).

In 1924 Louis de Broglie put forward his research findings regarding Subatomic particles, in particular suggesting that electrons behave also like waves; hence the duality of wave-particle aspects of the electrons.

In 1925 **Wolfgang Pauli** presents his 'exclusive principle', which is associated with the existence of electrons and their quantum state.

In 1925 **Werner Heisenberg** develops a 'self-consistent theory' of Quantum physics. It was the first in its kind and was later known as Matrix mechanics.

In 1925 **Paul Dirac presents** his 'operator theory'. This was a new version of quantum theory.

In 1926 **Max Born** put forward the concept of 'Quantum Probability'.

In 1926 **Enrico Fermi and Paul Dirac** formulate the rules that govern the behaviour of Fermions. After having done their calculations etc, they concluded that in the Universe the total number of Fermions is the same, suggesting that this was determined by the laws that existed in the Big Bang.

In 1926 **Erwin Schrodinger** publishes his wave mechanics equations, which are not only significant but also of great importance in quantum mechanics.

In 1926 **Paul Dirac** points out that Schrodinger's wave mechanics and Heisenberg's Matrix mechanics are relevant to his Operator theory.

In 1926 the name 'Photon' is given by Gilbert Lewis to the particle-like quantities of light introduced by Albert Einstein in 1905.

In 1926 **Heisenberg** discovers the 'Quantum Uncertainty', and he formulates his 'Uncertainty Principle', which may be stated thus,

"We can not measure the position and momentum of a quantum
  entity, say an electron, precisely

and at the same time. We can,
however, measure one of these
with accuracy, but we can not take,
at the same time, an accurate
measurement of the other one;
there will always be an element
of 'uncertainty' hence the
term 'Uncertainty Principle'."

In 1928 Paul Dirac developed the first 'Relativistic Quantum Mechanics'. This means he constructed a special kind of quantum mechanics, which incorporates in itself some of the important requirements of Einstein's Special Theory of Relativity.

In 1928 Bohr introduces his *Complementarity Principle*.

In 1929 Bohr presents his *Copenhagen Interpretation* of the Quantum World. He promoted its significance in quantum mechanics.

In 1929 Heisenberg and Wolfgang Pauli present their Lagrangian Model of quantum theory.

In 1930 Einstein and his colleague, Rosen, 'built their bridge' and called it 'Einstein-Rosen Bridge'.

In 1930 Wolfgang Pauli discovered, identified, the existence of the particle which later was called 'Neutrino'.

publishes 'The Principle of Quantum Mechanics'. This was the first in its kind.

In 1932 Carl Anderson becomes the discoverer of the Positron.

In 1934 Louis de Broglie introduces the term 'antiparticle'.

In 1935 *The EPR Paradox* is introduced by three people one of whom was Einstein. This states some of the paradoxical features of quantum physics.

The decade of the 1940s

In the 1940s, probably, the most significant achievement in terms of quantum physics was the development of *Quantum Electrodynamics (QED)*. *QED was developed by the end of the 1940s by three distinguished physicists, Julian Schwinger, American, Sin Itriro Tomanaga, Japanese and Richard Feynman, arguably the greatest British physicist of the 20th century.* (What about Paul Dirac? One of the most eminent Quantum Pioneers. I should rate him as one of the ten greatest physicists of the 20th century).

Although it was developed, as I said, by the end of the 1940s, Quantum Electrodynamics' foundation was laid by Paul Dirac in 1926; he also described the working of it by using the required mathematical equations, which he formulated. But what is it --- - what QED is about? To the best of my little knowledge, QED is about Quantum theory of the electronically charged particles with magnetic fields; this takes place by means of Einstein's light particles which later were called 'photons'. Among quantum entities such interaction also goes on with one another. The explanation of QED involves some highly elaborate mathematical calculations and certain complex equations. QED in truth (and in short) is very much a 'relativistic Theory'; central to all its mathematical procedures and its equations is some of the 'main ingredients' of Einstein's *Special Theory*.

The formulation of QED was followed by the meeting of two well-known quantum theorists, Enrico Fermi, Italian, and Hans Bethe, German (the latter later went to America where he lived the rest of his life). Hans Beth met Richard Feynman in America. He encouraged Feynman to work with him on a certain project; this was after the War. Enrico Fermi was truly a 'quantum genius', and there are many who believe, probably, he was the greatest quantum physicist. Today we have a few things after his name: 'Fermi', a unit of (wave-length), 'Fermi-Dirac statistics' which refers to certain statistical calculations and equations that are related to the behaviour of some of the quantum particles, 'Fermions', which are the particles that follow Fermi-Dirac statistical laws, and there is also 'Fermi constant' and so on.

The decade of the 1950s

In the decade 1950s still some of the quantum activities among a number of eminent physicists, including David Bohm and Richard Feynman, were going on. Some of these were significant and original concepts and quantum ideas.

The decade of the 1960s

This so-called 'permissive decade' was also 'scientifically permissive' by which I mean a great deal of activities and scientific events (perhaps more than any other decade) took place. A number of scientific theories, concepts, and new ideas were developed or put forward. And of course such innovations and discoveries, culminated in 1969 - man's landing on the Moon.

I am going to list roughly one third of the scientific events of the 1960s. These are not (necessarily) the most important ones, but they are, in my opinion, either directly or indirectly related (or at least have some relevance) to quantum mechanics.

In 1962 Richard Feynman, drawing upon his Quantum Electrodynamics, developed the concept of 'Quantum Gravity'.

In 1962 the second Neutrino was identified (or discovered).

In 1964 John Bell, based on his research findings, defends David Bohm's 'variation' on the EPR, which was put forward in 1952.

In 1964 **Gell-Mann and George Zweig** independently introduced the concept of *Quarks*.

In 1966 **Fred Hoyle, Bob Wagoner and Willy Fowler** put forward a new concept for a *better understanding of Big Bang Nucleosynthesis*.

In 1966 **John Bell** puts forward his scientific argument for the defence of David Bohm's 'hidden variables' theory and concludes that Von Neumann's proof, which played the important role in discrediting the theory in question, was wrong and completely false. This greatly increases the possibility for the Hidden Variable to become, perhaps in the near future, an important model of the quantum world.

In 1967 **John Wheeler** was (becomes) the first physicist to use the term *'Black Hole'* in a scientific (astronomical) context.

In 1967 **Andre Sakharov** in his scientific paper shows that there is (maybe) a very small asymmetry in the Laws of Physics. This has certain implications for the theory of Big Bang.

In 1969 Human beings landed on the moon! Unbelievable!

The decade of the 1970s

This was mostly the decade of space exploration and astrophysical and astronomical activities. It was the decade of Skylab Missions and of the launching of Space Probes by NASA, called the 'Voyagers'. In fact it was a highly busy decade for NASA.

In the decades of the 1980s and 1990s new ideas, fresh concepts, and interesting theories were put forward. Such activities and achievements are still going on, and will continue throughout the 21st century, and even beyond.

12

The Power of Imagination

And

Therapeutic Imagery

In the past three decades or so among the (psycho) therapists there has been an increasing interest in using visualization as part of their therapeutic works. The application of therapeutic imagery, visualization, and the belief in its beneficial effects (especially on the attainment to 'self-development' and also higher levels of consciousness) can be traced back to ancient times. In Hinduism the practice of visualization has always been used as part of certain types of meditation. In some schools and certain branches of Buddhism imagery has been in practice as part of their esoteric rituals. In Tibetan Buddhism, for example, imagery plays an important part in their spiritual disciplines including some of their meditations and especially in their healing rituals.

Before going any further let me say that once one of my psychotherapy tutors said to me that I should never use the word 'visualization' in a therapeutic context, but instead, I should use the word 'imagery'; also, I was told not to say 'to visualize' but instead to say, 'to image'. When I asked him the reason, he gave me a reasonable answer by explaining the subtle difference between them, especially therapeutically. I have since tried to follow his advice, however, every now and then in therapeutic contexts I also say 'to visualize' and 'visualization'-----Sorry Sir!

There are some people who say that they cannot image. I believe, in fact I am sure, everyone can image (visualize), though some are better at it than others; it is all a matter of degrees. There are 'good visualizers' and there are 'poor visualizers'. However, the regular practice of some specially-devised techniques can improve, often to a large extent, the person's ability to image. Indeed, there are special exercises that can increase ('enhance'), usually considerably, the power of imagination----hence, the increase in the ability to image during therapy. It is therefore important for a psychotherapist who uses therapeutic imagery as part of his or her work to have the knowledge and skill of devising such exercises for improving the power (capacity) of the imagination of the person (the patient) concerned. At this point let me define some of the important terms that are used in this essay.

To imagine, to experience something that is non-existent to the senses; to experience (an object) in the mind.

Imagination, the process or the act of imagining.

To image, to see, in therapeutic context, with 'mind's eye'; to picture in mind; to visualize.

Imagery, the act of imaging; visual imagination.

The major imaginations are of visual, auditory, gustatory, tactile, and olfactory nature. And among these the most powerful, especially therapeutically, is visual imagination, or simply 'imagery'.

Imagination is an 'indispensable gift' conferred upon human being. Imagination in general is associated with abstract thinking and also is closely related to 'creative thinking'. Animals don't have this 'precious gift', the ability to imagine. If animals had such 'capacities', they would be able to innovate, to design and to create. *Imagination is the 'foundation' of creation; it is the 'corner-stone' of every innovation and of every creative activity. Any good artist will testify to the fact that without imagination (artistic) creativity would be impossible.*

If we use (regularly and with a positive attitude) our imaginations, particularly through imagery for the purpose of our personal growth and self-development we shall, most probably, succeed in attaining our aims. *The power of imagination is central to the power of mind. And incorporated in powerful minds are powerful imaginations. Powerful imaginations lead to what is sometimes referred to as 'imaginativeness'* ---- *the 'seat' of creativity and innovations.*

In the early part of the 20ᵗʰ century there was a French psychologist, *Emile Coue,* (well- known for his celebrated statement: **"Every day in every way I get better and better"**), was convinced (and so am I) that *imagination is more powerful than will.* The reason for this, in my view, is that imaginations involve elements that are mostly of emotional nature.

Imagination in general and imagery in particular, in the context of therapy or healing, are the most powerful 'tools' in enabling the mind of the patient to take an active part in the treatment of her disease. This activates the patient's healing resources including the subtle energies, thus giving rise to healing and restoring health in her. This (process) I have called *'Psycho-healing'* which is the *'active participation of the patient, or rather the patient's mind, in the treatment of her disease.* I will talk of this more later.

Imaginations can, and often unknowingly do, play important part in our lives. Great men and women, who have succeeded in self-transformation, have had access to the inner resources of their minds and in particular they have been able to utilize the power of their imaginations. Great achievers, in general, are imaginative people. They usually imagine vividly their ambitions and image intensely their prospective achievements.

To many people, sadly, the word 'imagination' means a useless activity of the mind. They use the word in question in a derogatory way. For example, 'You are just imagining'; 'He is good at imagining and that is all'. To these people 'imagination' is nothing more than a useless mental activity of the idle people. They are ignorant of the fact that the world owes its tremendous advances, innovations, creativities and cultural masterpieces to the great people's imaginations. Indeed it is the 'imaginativeness' of the person that innovates and creates. *Every creative activity is preceded by the imaginations of its creator.*

Through the regular practice of the 'proper exercises', as has been said already, the ability to imagine can be improved considerably. There are special techniques that, if practised over a period of time, can enhance the effect of therapeutic imagery and increase the power of imagination. Such techniques and exercises all come under the 'umbrella' of *the training of the mind*, and there is no particular (a single) method for such training but there are different practices ---- various methods and techniques.

Having said all this, I am going to 'reveal' an *imaginative secret!* And that is the truth about imagination: *underlying the imaginations or rather the power of imagination is the controlling of the thoughts, and this also applies to the practice of imagery in any therapeutic context.*

'Controlled thoughts' play the crucial role in rendering therapeutic imagery so effective. Thinking in a controlled manner (controlled thought) and the act of imagining, in truth, are inseparable. Put another way, in any act of imagining, imagination, the controlling of the thought is present whether it is the picturing of a red apple in the mind, or imagining oneself standing on Westminster bridge, or experiencing, in the mind, climbing a mountain. In all these 'imaginational experiences' the imaginer aims at controlling his or her thoughts on the object (or act) of imagination.

Therefore it is greatly important to learn to have control over our thoughts; this is central to the training of mind----the act of having control over our thoughts. In order to achieve this, it is essential to learn how to avoid 'the uninvited thoughts' entering the mind. Mystics of any belief system since ancient times have known this 'priceless secret' as part of their disciplines on the 'mystical path'----*to think what I want to think*. **Great Systematizer of Yoga, Patanjali** (who lived about two thousand years ago) *says, 'Yoga is the control of thought-waves in mind'*. Indeed, central to yoga meditation is the concentration (one-pointed focus) of the thoughts.

We can see, from what has been said so far, that imagination is the imaged (controlled) thoughts in mind. And I go further and conclude: *Consciously controlled and purposefully imaged thoughts of positive or healing nature are powerful and highly effective whether it is in a therapeutic context or in daily life*. The 'imagined thoughts' in therapy, therapeutic imagery, activate the healing resources and especially revitalize the bio energies, the subtle energies, in the person. And all these together enhance the capacity for self-healing, strengthen the ability to persevere, and sustain a high confidence to carry on the 'healing works'.

Our imaginations are our private experiences and we are free to indulge in them almost whenever and wherever we want. We must therefore learn to make the best of such 'privacy and freedom'; for example to learn to think constructively and to imagine creatively. This is a crucial part of the training of mind, and it helps immensely to acquire powerful imagination, which often gives rise to imaginative ability, followed by the creative capacity in action. Put in a nutshell, ***creative activity is creative imagination put into practice.***

It is noteworthy to know that 'healing' is a creative act. To bring about healing whether in oneself or in others, or to engage in therapeutic work is an activity related to creativity. In fact psychotherapy is an 'art', a creative art. It is an art that involves (as part of therapeutic experience) insight, intuition, empathy, also interaction with the patient; inspiration, too, may play a part but this depends on the therapist as a person. And the application of imagination in therapy (I have in mind mainly therapeutic imagery) is very much an art requiring a reasonable knowledge of, and at least some experiential skill in, this therapeutically interesting area.

The psychotherapist ought to make sure that the content of imagery is suitable for the patient; it must be closely associated with her problem(s) and it has to be relevant to her situation. The therapist should remember that the patient is a unique individual with a unique set of life experiences and with a problem that is unique to the patient. Therefore, the content of 'therapeutic imagery' has to be so devised as to be uniquely suitable for the patient and 'tailored' to her problem or the situation that she is in. Of course, here I am speaking of 'specially-devised therapeutic imagery'. However, there is also plenty of material from which the therapist can prepare therapeutic imagery that may apply to everyone (to more than 95% of people)----'general therapeutic imagery'.

The more a therapist knows about the patient (her aims, ambitions, hopes, likes and dislikes etc.) the better, more suitable, therapeutic imagery he can devise for the patient. That is why every therapeutic session should be preceded by ten or fifteen minutes of counselling in order for the therapist to have a better understanding of her problems and a clearer picture of the situation that she is in. And these will enable the therapist to devise the content of the therapeutic imagery more confidently. This means he will know what kind of 'imagery material' to choose and skilfully to include in the content of the therapeutic imagery.

Human mind is not as straightforward as it might seem to be, particularly when it comes to therapy. The mind doesn't mind to be given encouragement in order to better itself, but it does mind to be coerced or insisted upon to transform itself. Human mind by its very nature likes to find the truth for itself and by itself; it likes to unfold the enfolded by itself. Within a reasonable context the mind welcomes some advice or guidance, but it doesn't want, especially in therapeutic context, to be pushed into doing things that it doesn't want to do. Those who specialize in this therapeutic area have noticed that the mind is by far more receptive to indirect statements, implicit remarks, and embedded directions than direct statements, explicit remarks and straightforward suggestions. The reason for this, in my view, is that indirectness and implicitness etc give rise to curiosity in human mind, and very often curiosity gives rise to imaginations, which in turn may intensify the curiosity.

Now we can see why some of the greatest minds, particularly the Mystics, Enlightened individuals and some of the great philosophic thinkers and metaphysical poets since ancient times have, throughout their teachings, used parables, proverbs metaphors, implicit statements and allegories. They knew the 'secret' of arousing curiosity in human beings, and also of leading their minds to imaginations so that all these may give rise to finding 'the truth' for themselves. These great human beings knew that directness and explicitness seldom entice the minds to ponder curiously and almost never to think imaginatively. The Illumined men and women are well aware of the 'secret power' of parables and allegories; these imaginative individuals know how to convey in the most effective manner their teachings or messages to people. Indeed, they know well about the 'nature of human mind'. The allegorical story in The New Testament, known as the 'Parable of The Prodigal Son' told by Christ, is of Highest Order; it is said to be the most beautiful (meaningfully beautiful) shortest story that there has ever been. Lord Jesus has Masterfully embedded so allegorically meaningful words in this spiritually sublime parable. (Luke, 15: 11-32)

The embedding of implicit words is also used in therapeutic imagery; experienced and skilful imagery therapists can do this properly. The therapist every now and then embeds in therapeutic imagery certain metaphors or implicit suggestions or allegorical statements all of therapeutic nature. Indeed subtle comments, metaphors, parables allegorical stories stimulate our minds: they make us think, and they arouse our curiosities and our imaginations so that we may find the truth or to grasp some meaning behind them. Here I must say a few words about the late professor **Milton Erickson** who was a superb master in using parables, metaphors, indirect suggestions, all implicitly of therapeutic nature intended to bring about healing in the patient. Though a psychiatrist, Erickson practised psychotherapy of his own. He innovated a kind of psychotherapy, Ericksonian psychotherapy, which involves skilfully designed approaches, and these involve special skills and techniques that require proper training. He was a 'creative psychotherapist' in the true sense. I watched this *therapeutically imaginative* man on the video administering his psychotherapy. How original he was in his profession! When talking to the patient, he would embed healing words and therapeutic phrases, all in an indirect and implicit way. It is mainly such skills and original techniques of therapeutic nature that have made his therapy so effective. He developed his therapeutic method mainly through his understanding the nature of human mind, including its (mind's) peculiarities and idiosyncrasies, particularly during psychotherapy. Moreover, Erickson realized how immense and imaginatively great the human mind is. In fact, underlying Ericksonian psychotherapy is the attempt to stimulate mind's healing resources mainly by its therapeutically imaginative curiosity, Erickson knew well about the power of imagination especially in therapeutic setting. Indeed, I attest this. From my studies and research in such areas, and also from my both professional and personal experiences I have realized the astonishing power of human mind. Let me try, only briefly, to summarize the greatness and the imaginativeness of the mind, including its creative capacity in therapeutic/ healing works.

In the depth of the ocean of our imaginations lie our creative potentialities and unrealised capacities. By diving into our potentially imaginative ocean we can discover our creative world and our dormant talents; and in the process we also get to know the nature of our minds. Indeed, we find that despite of being a strange, paradoxical and idiosyncratic 'stuff', human mind is truly and marvellously grand. Such 'Greatness' of Mind lies mainly in its incredible resources: a) its creative capacity ('imaginativeness'), b) its healing power ('therapeutic imaginations'). These two are not unrelated. Healing, in its own way, is a creative ability, and therapeutic skills are a kind of creative skills. Imaginativeness is the creative capacity and the seat of innovations in human being. This 'precious gift', power of imagination, has to be nurtured and needs to be cultivated; however, some people have to work much harder than others on such 'cultivation'. This is because we all are not (cannot be) creators on the same scale; creative capacity varies among human beings.

What is greatly important to remember is that imagination is potentially powerful, moreover, thoughts of healing nature when consciously controlled and purposefully imaged are beyond any doubt immensely forceful and positively powerful. Therefore, let us train our minds, especially in imagining our ambitions vividly and imaging our aims or goals intensely. And if we practise these techniques regularly, patiently and wholeheartedly, then the fruits of our efforts and the successes that we expect are not just a possibility, but even more than a probability. And we ought to share the fruits of such successes with our fellow human beings ---- to let the society that we live in, and perhaps also the world that we inhabit, as far as feasible, to benefit from them.

13

A Journey into
The Realm of Mystical
Meditation

This essay is on my journey into the 'realm of meditation'. It is the 'product' of my personal experiences of meditation, and also my studies and research into different systems and various types of meditation and related areas including mystical schools of different belief systems that I did many years ago.

Before going further let me make clear one thing which may be important to some of the readers of this essay to know. The reason that in the title I have chosen the phrase, 'realm of mystical meditation', instead of simply 'realm of meditation' is to emphasize that the subject matter that I am going to discuss here is purely spiritual, and the systematic practice of meditation considered in this essay is solely for spiritual purpose, as has been for thousands of years in the East. Indeed, in the Eastern World meditation is practised for raising the 'Level of Consciousness', in order to enter the '*Higher Realm*' and to get closer and closer to the 'Spirit'. However, this is not quite so in the West. Great majority (yes indeed, great majority) of people in the Western World take up the practice of meditation for reducing their stress and alleviating their anxieties; in short, to acquire some peace of mind. There is nothing wrong with such 'meditational ambitions'; however, in the East these are regarded as (and they are) the 'by-products' of the regular practice of meditation. The purpose of meditation is and should be (primarily must be) a spiritual one. Yes, it is good to know what 'Meditation' is about; it is purely a 'Spiritual Act'. Now we can see why I have chosen the qualifying adjective 'Mystical' for the noun 'Meditation'. Apart from reminding us of *mystical eternity* and the *mystics* (The Awakened, The Enlightened, or Holy people), the word *mystical* also often refers to the highest form of spirituality, and it is mainly in such context, mystical context, that I discuss Meditation in this essay and throughout my 'journey'.

Now it is time for me to set out on my journey in the *Realm of Mystical Meditation*. It is an adventurous journey and a formidable task, which, I am sure, I will enjoy ------- I hope so anyway.

Meditation opens the gate to our inner silence, and our inner silence is the entrance to the mystical realm where the practising aspirant may, one day, over a period of time, experience some transitory 'inner sparks of divine eternity'. And let us remember that in the depth of our innermost silence there is a spark of *Mystical Eternity*.

Meditation is also the Path to Wisdom. Such wisdom together with our spiritual growth progressively enables us to acquire self-knowledge and obtain insights into ourselves----into our personal problems, even into our inner conflicts; we gradually find the courage of confronting our 'shadows' that we always pretended they had nothing to do with us----as if they were not and had never been part of us.

Silence, particularly the silence of meditation, is not passive but dynamic even sometimes terrifying. "The eternal silence of these infinite spaces frightens me," said Pascal. In the Silence of meditation we become face to face with ourselves, and we see, if we dare to look at, our dark sides. Most people, but not all, who cannot meditate are among those people who cannot stand silence particularly their *Inner Silence*. But some people feel at home in their inner silence.

The Silence of Meditation

It is in *Silence* that we sense our belonging,
It is in stillness that we feel our longing,
It is in stillness and silence that we
Discover our longing for belonging.
The farther we are the more we long,
The closer we are the more we belong.
Longing, in a way, is a cry for affection,
Belonging, somehow, denotes a connection.
The higher is the flame of longing
The deeper is the sense of belonging.
Meditaters meditate for they know in *Silence*
Meditation takes them to where they belong--- Home.

In meditation the closer we are to Home, the more we feel at home; hence, greater becomes our knowledge of our true nature. And such knowledge and insights into ourselves render us capable of understanding our 'true self'. (In such context I prefer not to use the word 'self' as plural) Self has never been a 'favourite thing' among psychologists with the exception of those who adhere to Humanistic Psychology. This 'psychological ostracism' of 'self' has mainly, but not only, been due to the status of 'mind' and more recently of 'consciousness' in psychology. But then let us not forget that until thirty or forty years ago in psychology even the word 'mind' was used seldom and 'consciousness' hardly. Compare this with today's great enthusiasm among some psychologists for doing research into 'Consciousness'.

'Consciousness' is one of the two psychological terms (words) that often are used in relation to Meditation. The other one is *'thought'*. Consciousness is closely related to the practice of meditation. In fact the very 'act of meditation' aims at raising the *Level of Consciousness* (whether the mediator knows this or not matters not). Furthermore, the practice of meditation is a *spiritual act*. In fact meditation is 'prayer' in its broadest sense (not 'prayer' in its strict, theistic, sense but in a broad sense). Let me define (my own definition) 'prayer' in its broadest sense: "Any attempt used by a person for the purpose of raising his or her "Level of Consciousness" is prayer". This means the Buddhist Monk who gets up at four o'clock in the morning and starts doing his zazen meditation, he is praying; the Icelandic student who, in her London bed-sitter, is meditating, she is praying; the Sufi who, in the Khaneqah, is lovingly contemplating the Highest he is praying; and so is the Yogi when he has, in lotus position, been concentrating for nearly an hour on the beautiful flower in front of him. In each one of these instances the attempt is to raise the Consciousness to a Higher Level (or, to the Spirit) ------- to experience The Presence. So, we can see how closely Consciousness is related to the act of Meditation. And so is 'Thought' which now I am going to talk about briefly.

In the definitions and descriptions of different systems and types of meditation the term *Thought* or *Thoughts,* is probably the most used one. The word is used in meditational contexts such as 'observing the *thoughts'*, 'quieting the *thoughts'*, 'attending to (being attentive to) the *thoughts'*, 'meditating or concentrating on a *thought'*, 'the absence of the *thought'*, 'controlling the *thoughts' and so on.* Patanjali, the systematizer of Yoga, defines *Yoga* (in meditational context) as **"The controlling of the thought-wave"**.

Spirit and Consciousness

Though not the same, Spirit and Consciousness are on the same spectrum; on one end is consciousness, and on the other end is 'spirit'. Between them there is no clear-cut line; it is a matter of degrees. Spirit is the refined, pure and unadulterated consciousness; it lacks the knowledge of evil. The 'spiritual definition' of the word 'innocent' is "The state of lacking the knowledge of evil". I may not be (and I hope I am not!) an evil person, but I (and my consciousness) do not lack the knowledge of evil. Now we can see why we say a child is innocent; not because he or she is not evil (we all know that children are not evil), but because a child's mind, child's consciousness, lacks the 'knowledge of evil'. Children's consciousness is (or is very close to) 'Spirit'. That is why we are told unless we are like children (pure at heart and innocent in mind) we cannot enter the 'Kingdom of Heaven' (and we know who said that). We therefore can deduce from what has been said: that the closer is our consciousness to the Spirit, the higher is 'the level' of our consciousness. When we practise (practise systematically) meditation our consciousness gets closer and closer to the Spirit. Therefore, the phrase, 'The level of Consciousness' denotes 'the spiritualization of the consciousness'. For example, if I say 'Tom has a higher level of consciousness than John has', I simply mean, or imply, that Tom's consciousness is more 'spiritualised' than John's consciousness is.

'Spiritualization' in such context is a tender process of 'whispering' some loving words in the ear of the meditater who also, at the same time, experiences, or feels, the tenderness of the 'magic of Love'.

> **Whisper to me like a Lover,**
> **For tenderness is rare in the world.**
> **It is difficult to convey the magic of Love**
> **To those who are made of dust.**

Consciousness is closely related and Thought is very much relevant to Meditation ------- an inseparable relationship. 'Word', too, plays a highly important role in such affinity. Some say 'Word' is the 'most powerful thing'. In certain types of meditation the repetition of a word, (a good, positive, spiritually significant, or a beautiful word) is a common practice; such meditation is referred to as 'Mantra Meditation'. Word is the 'product' of a thought. (Some would say the reverse is true.) St John's Gospel, which is a Mystical Book, begins with,

"In the beginning was the Word, and the Word was with God, and the Word was God."

Above statement is "The Mystery of The Mystical Eternity" --- truly the 'Cosmic Mystery'. This reminds me of a special personal experience; I will tell it to the readers of this essay. Many years ago I had the privilege of having an audience with a Sufi Master (Peer). I will always cherish that 'spiritual occasion' which stands out in *the field of my memories.* The audience lasted nearly an hour and throughout we conversed in the beautiful language of Farsi, Persian language. Farsi is greatly rich in literature, and when it comes to Poets and Poetry and the sublimeness of the Poems, it is second to none. (I am not a Persian)

The Great Sufi had already been told a few things about me. He knew I am a Christian, and also he knew that I have a great interest in mysticism, mystical spirituality, of different faiths particularly in Sufism. Wise master, a Moslem of course, in order to make me 'feel at home', began speaking to me about Jesus Christ and His 'Place' in Sufism especially among Sufi Poets. I said that I know that; then I said especially Rumi in whose poetry Jesus is an often-mentioned name. Then one of the first questions the Peer (Sufi master) asked me was if I had read the "Enjeel e Yohanna" (meaning the 'New Testament of John'--------St John's Gospel). I told him that I had read it many times, and that I had read it in the language spoken by Christ, in Aramaic. He looked at me somewhat with surprise, and then he went 'Bah-Bah! Bah-Bah!'' (An exclamatory utterance among the Persians, such as *Well Done! That is Wonderful! fantastic!* etc) The Peer, then said that St John's Gospel is one of the great mystical books. We agreed that it is truly a mystical book. We both also agreed that this is evident especially in the last few chapters (of the Gospel in question) which include so many 'mystical statements' made by Jesus, all referring to 'Oneness', 'Union with God', 'Mystical Eternity' -----e.g. '*...they may be one as we are*', '*...and I in them*', '*... loved me before the foundation of the world*', '*.....they may be one, even as we are one*', '*......all mine are thine, and thine are mine*', '*... I and Father are one*', '*I am in the Father and Father is in me...*', '*....abide in me as I abide in you*', ... '*On that day you shall know that I am in my Father, and ye in me, and I in you.*' .... and so on.

'Words' among spiritually developed individuals in general and mystics in particular matter a great deal; 'words' are treated with special care and some words even with reverence. There is a 'Special Love' for words among the poets, in particular mystical poets.

Word is more than 'just a word'. In the Aramaic New Testament St John uses the word '**Milta**' for *Word*. "**In the beginning was Milta, Milta was with God and God was that Milta**". One of the meanings of the word 'Milta' is 'word' but also it has other meanings, such as 'Manifestation', 'Spoken Word', 'Verb', 'Substance', 'Foundation' and probably more. 'Milta' still remains both spiritually and philologically somewhat a 'mystery, even among the speakers of Aramaic. Here it is noteworthy to mention that Aramaic is unique in metaphors, similes, parables and proverbs. It is a rich language and distinctive of being holistic, mystical, and especially it is a healing-centred language. There are so many words in daily conversation all derived from the word 'Healing'. Aramaic is a meditative and contemplative language.. Aramaic speakers love their language and to them words matter a great deal. 'Word' is Eternal, and the words of spiritual nature don't die away. '*Heaven and Earth shall pass away, but The Word that proceedeth from the mouth of God shall not pass away*".

It is beautiful in the *'Inner Silence'* to meditate, concentrate or contemplate (for five or ten minutes as part of the 'proper daily meditation') on a beautiful word, especially on a *Word of Spiritual Significance,* or on a *Phrase of Mystical Nature.* Such practice calms the Mind and uplifts the Soul. And if this is done regularly over a period of time, the person gradually will notice that 'things' are beginning to happen. There are different systems and various types of meditation. It is good to choose the kind of meditation that suits the person's personality and fits his or her psycho-physical (chiefly psychological) make-up. However, one can choose any type of meditation; the person gradually (sooner or later) will get used to it and will most probably practise it well.

The systematic practice of meditation progressively enables us to acquire insights into our inner Self even our true nature. We come to the realization that human beings are not 'walking bodies' but 'multi-dimensional beings'------- beings of different dimensions and various levels, notably of Mind body and Soul. Meditation progressively renders us aware of our 'Self'. We become 'aware of our Body and of our Mind and particularly of their interrelation', and this I call *'Self-Awareness'.* Hidden in the *Self* of each and every person is **The Eternal Spark of Divinity** from which we can draw strength and courage even progressively we may be able to attain **'the Peace that passeth understanding'**. This kind of Peace is the Peace of the Spirit. Human consciousness is part of The Universal Consciousness and is capable of being One with *The Spirit*, which is boundless and without boundaries. The great Zen Master Huang Pu says,

> **"All the Buddhas and all sentient beings are nothing but
> One Spirit, beside which nothing exists. This Spirit,
> which is without beginning, is unborn and
> indestructible........ It transcends all limits,
> measures, names, traces and comparisons.
> Only Awake to the One Spirit."**

The journey into the 'Spiritual realm' over a period of time gradually, step by step, renders us aware of *Mystical Eternity*. In this and with this 'realm' one begins to fall in love with the 'Infinite', because one is beginning *to be in touch with the Infinite Love. **It is beautiful to fall in love with the beauty of a woman, but to be in Love with the 'Beauty' of God is beyond any beauty; it is indescribable.*** The mystical poet of mystical poets of all times, Mowlana Rumi (known in the West simply as Rumi) calls this indescribable Love as 'tongueless Love'. By 'tongueless' he refers to the inability of the tongue to explain or to describe Love. And when he comes to write about Love, to explain it in writing, 'something strange' happens to his pen. I will let Rumi, The Persian, to say it:

> **Whatever I say in exposition and explanation of Love,**
> **When it comes to Love I am ashamed of that Explanation.**
> **Although the commentary of the tongue makes clear,**
> **Yet, tongueless Love is clearer.**
> **When the pen was making haste in writing,**
> **It split up itself as soon as it came to Love.**
> **For expounding Love the intellect lay down helplessly**
> **Like an ass in the mire.**
> **It was Love alone that uttered the explanation**
> **Of Love and Lovehood.**

It is somewhat a 'poetical crime' to read the translation of Rumi poems. They should be read (if one wants to experience their 'sublime beauty') in the beautiful language of Rumi, Farsi (Persian language). The translation of poetry in particular the mystical poems corrupts the essence and trivializes the profundity of the poems that were written in the original language. The translation can never express the beauty nor can it present the sublimeness of such poems; and what about the depth of their meanings?

Sufis are in Love with God. **Love and being in Love with God is for them all that matters.** Rumi was desparately in Love with Love. One night he took the courage of asking Love, to define Herself.

He wanted to find a 'definition' of Love that could fit, or match, his mystical experience of Love or at least his mystical identity. He says,

> **One night I asked Love, say truly who**
> **Are you?**
> **She replied: I am everlasting life,**
> **The succession of happy Life.**

So, Love replies, 'I am everlasting life' ----- claiming Eternity. Christ Jesus, too, by breaking the grammatical rule in the statement, 'Before Abraham was I am', claimed Eternity, or rather Divinity. He knew he is Divine. *And where the Divine Presence is, the rules of grammar run away: past tense and future tense melt away and syntax looses its meaning. and all that is left is the present tense of The Presence ---- Eternal.*

However, the mystically wise Rum was cautious in the following poem not to break the rule of grammar, because he knew that he was not divine, but only the experiencer of 'transitory sparks of divinity', and of 'momentary mystical eternity'. Notice how careful he is not to claim divinity --- he says (as he journeys into the realm of meditation),

> **The universe wasn't there I was** [*not 'I am'*]
> **Adam wasn't there and I was.**
>
> ....
> **The universe got light from me,**
> **Adam took his form from me.**

Sufism existed in Iran two hundred years (probably more) before Islam. This may come as a surprise to many people both Moslems and non-Moslems; but it is true. Later, Islamic theologians and Moslem Scholars including Al-Ghazali (1058-1111), born in Iran, succeeded in making Sufism acceptable as the mystical school (or branch) of Islam; of course this took place as a gradual process. Today Sufism is known as the mystical branch of Islam; however, Jesus, as the Sufi Master put it, has a special place among the Sufis.

Jesus says, if you don't loose your life, you cannot gain your life. Sufis say, if you don't annihilate your lower self (referring to the stage of Fana) you cannot attain to your Higher Self. Jesus says, be perfect even as your Heavenly Father is Perfect. Sufis say the aim of every seeker of The Truth must be Perfection (Kamahl)

On the Path, the Mystics of any religion or belief system have a 'burning desire' to get closer and closer to the Divine or Sacred Entity, and finally, to become in Union with the Ultimate Reality. And this is the 'enthronement' of the Soul in the 'mystical eternity'. Such Mystical Union is enveloped in Light, Shining inextinguishable Light. Before the Creation this Light is. It is Alap and Taw, the Beginning and the End; yet before the Beginning this Light is.

Meditation is an Avenue capable of leading the Meditater to Mystical Eternity; Meditation on the Mystical Path aims at Enlightenment. Meditation, if practised in a systematic way, and with Total Commitment, and with Healing thoughts and purity of Heart is The Highway to Splendour of the Truth and to the burning Fire of Love, even God.

> **The time has come to turn your heart**
> **Into a temple of fire.**
> **Your essence is gold hidden in dust.**
> **To reveal its Splendour you need**
> **To burn in the Fire of Love.**

*As a 'mystical aspirant' (or 'student of mysticism'), I have come to the end of my **Journey into the Realm of Mystical Meditation**, an 'endless' journey; and it is also an endless subject matter which I have dealt with in something like four thousand words. I feel a bit tired, but 'beautifully tired'. The journey has not been an easy one, yet I have enjoyed every* *second of it. The Path, at least in part, has been an uphill task; yet, I have also experienced beautiful moments.* I wish the Reader of this essay 'Beautiful Moments' too, especially in the 'realm of mystical eternity'. ***All The Blessings* ----- *Shlama.***

14

Self-Healing;
Reversing The Process

The term 'Self-Healing' refers to healing oneself, or to one's natural capacity for bringing about healing and restoring health in oneself, or it may also refer to the method, technique or any process that one engages in for effecting health and a sense of well-being in oneself. Everyone who practises self-healing has his or her own way of healing themselves, usually depending on the kind of the belief systems, philosophies, or the concepts of healing that they hold or follow.

Self-healing does not have to be practised only when one is ill or suffering from a disease; it should be part of one's way of life. The regular practice of any method or technique of (self-) healing nature is highly beneficial and therapeutically effective at all levels. Furthermore, it is perhaps the best way of preventing oneself from being affected by illness or a disease. The success in self-healing depends on many factors. The attitude of the person who practises self-healing is paramount. The person's commitment to embarking on such task matters a lot. Self-healing is an important and serious task.

It is an acknowledged fact, based on ample psychosomatic research findings, that in the lives of many people certain psychological processes such as unhealthy emotions, despair, self-deprecating feelings misconceptions and negative attitudes to life, and all kinds of psychosocial problems mostly of emotional nature, have gradually rendered them susceptible to illnesses and even to serious diseases including cancer.

Such susceptibility, chiefly of psychosomatic nature, can be remedied and the process of illness may be reversed by the regular practice of self-healing. Success in self-healing, as has been said, depends on a number of factors. The first three most important factors essential to self-healing, are: a) the belief, 'the acceptance', that in human being there is a self-healing capacity---- a natural, an innate, transformative capacity; b) a 'strong desire' to get well and to feel a sense of well-being in oneself; c) a 'genuine trust' in the whole self-healing process, healing practices.

In self-healing we gradually replace, or rather 'reverse' the unhealthy conditions and the destructive elements that have played part in causing, or are about to precipitate, illness in us. In the process of being healed, our false beliefs, wrong attitudes, misconceptions give way to a constructive frame of mind and to a healthy attitude to life. This is a *Reversal Process*, or put technically, it is *'Psycho-Physiological Reversal Process'*; it starts with psychological reversal followed by physiological one. For example, when a cancer patient practises self-healing regularly the healing first starts ('touches' the patient) at psychological level, or to be more specific, the self-healing begins to heal the psychological processes that began to trigger (and then played an important part in the development of) the disease in her. The healing progressively reverses these processes, thus paving the way for the reversal of physiological processes in her. The reversal of such psycho- physiological process in the patient is the result of a 'successful self-healing' taking place in her.

Many years ago Doctor Carl Simonton, the well-known American oncologist, and his wife, a psychotherapist, in their book, *Getting Well Again,* wrote,

> **'Our purpose in writing this book is to show that the cycle of cancer development can be reversed. The pathways by which feelings can be translated into physiological condition leading to cancer growth can also be used to restore health.'**

Today a growing number of medical people are coming to the realization that patients, especially those with serious diseases, need healing or as I call it *psychological treatment.* I firmly believe (and I know this from my professional experience) that such treatment is highly important, even in many instances essential, in restoring health and a sense of well being in the patient. This kind of therapeutic approach enables the patient to take part in the treatment of her illness; indeed the patient's mind should (in fact 'must') be actively involved in the whole therapeutic process.

Human body is not (as the great French philosopher, Rene Descartes, believed) a 'physical object' obeying physical laws; this is part of what is known as 'Cartesian Concept' (Descartes' Mind-Body concept). Human body is a 'living thing'; it is a 'ground-field' of bioenergies. Furthermore, human body does not obey physical laws but it works in accordance with certain psychosomatic (psychophysiologic) interrelations and interactions and it follows, functionally, some psychobioenergetic patterns.

Cartesian concept eventually culminated in Newton's physics and Newtonian 'Model of Reality', which is a mechanistic one------a mechanistic model of the universe. And when it comes to 'matter and energy', Newton believed matter was 'solid object' and nothing more. However, this was not the case according to Einstein who saw matter not as solid object but as a form of energy. Einstein showed us that matter and energy are two sides of the same reality; even his famous equation, $E=mc^2$, denotes that mass and energy are convertible. Einstein's physics and particularly his concept of energy and matter shook the foundation of the classical (Newtonian) physics, and some years later Quantum Mechanics demolish it completely.

Einstein's *Model of Reality* is fascinating and his concept of 'matter and energy', scientifically, is (was at the time) a revolutionary one. Applying such 'model of reality' to human being, we get a picture of human being quite different from the one we get through Cartesian concept or Newtonian model of reality. We see human being as a complex network of energies. Indeed, human being is a complex network and a multidimensional being of subtle energies. Human body may be thought of as, and it is, a 'ground-field' of energies, the bioenergies, which nourish and 'animate' all the organs, parts and even every single cell of the body. The whole body, even at cellular level, is permeated with 'life'.

The subtle energy especially the major centres of energies, usually referred to as 'Chakras', are closely associated with consciousness (levels of consciousness). Now we can see the importance and the crucial role of the mind in self-healing, in activating the healing resources including the flow of the bioenergies in the body. The high influence of mind on body is an indisputable fact. Furthermore, there is a constant interaction between mind and brain and they function as 'one whole'; and within such 'holistic setting', it is the mind that is the 'primary'. This means between the two, mind and body (brain), it is the mind that by far has the higher, 'superior', influence; its influence over the body is tremendous. This comes under the topic referred to as *Psychosomatics*, which is the science of the interrelation and the interaction of mind and body, in particular, the influence of mind on body.

As I said earlier, in the lives of many people a series of stressful psychological and psychosocial processes mostly of emotional nature go on over a period of many years. Such stresses and strains gradually set up in them the susceptibility to being affected by illnesses, even at times by serious diseases. In Self-healing the mind gently heals itself, that is, it tenderly heals the emotional aspects of the mind that have (usually for too long) been through the stresses and strains of life. This is followed by the healing of the body, and finally in a progressive manner the healing of the whole person. Indeed, when we let our mind exert its influence positively and in a 'Healing Way' on us we are in fact letting our mental processes 'favourably' influence our psychophysiological functions, reactivating the healing resources and revitalizing the 'subtle energies' in us. In the process, we have succeeded (and *this is what* 'Self-healing' is about*) in reversing the cycle of our ill-health problems (illnesses or diseases) by effecting healing in ourselves----Replacing Illness with Wellness. And this 'wellness' is not the wellness of the body only but the wellness of the 'whole person'. Indeed, Self-healing in its true sense is a holistic process.*

There isn't, however, only one method of healing oneself or others, but many ways and different techniques of healing. Self-healing should be ('must be') made part of the working of the mind. I am going to introduce a simple *Self-healing method,* the systematic practice of which will also help the person concerned develop a **Self-healing mind**. This method is effective and, if practised regularly, is highly beneficial. I recommend it to everyone. For putting it together I have drawn on a few therapeutic techniques such as yoga, therapeutic imagery, and deep form of psychophysical relaxation.

A Self-healing Method

To be on the *'Path of Self-Healing'* is beautiful and self-assuring; also to embark on such a Path is a 'serious' task. Indeed, self-healing is a 'serious business'. It needs commitment and perseverance and it has to be done systematically. Twice a day (say, once in the morning and once in the afternoon) is the best; fifteen minutes or so for each session is reasonable. Later, gradually of course, this can be extended to half an hour per session.

Position. Sit on a straight-back chair. Sit in the front part of the chair. You must not lean on the back of the chair. Sit straight-----comfortably straight. Let your hands (the palms of your hands) rest on your knees. And let your feet rest flat on the ground / floor. Having positioned yourself comfortably, you are ready to begin.

1) Close your eyes gently. Attend to your breathing: notice your 'breathing in' and 'breathing out'. Do this, focusing on your breathing, for a few minutes. Don't change the pattern of your breathing. Let your breathing (in and out) take place normally, and you will notice that it is gradually and gently becoming rhythmic. (Incidentally don't worry about time---how many minutes you will do that or how many minutes you will do what. As you go on doing your healing exercises, you will gradually learn, 'intuitively', how much time, without knowing the time, you have spent or want to spend on doing 'a particular thing'. Time, the passage of time, will take care of itself.)

2) Now, imagine, and imagine vividly, it is a good day and you are sitting in pleasant surrounding, perhaps a favourite spot of yours: a beautiful garden, seaside or any place of your choice. Image (visualize) yourself looking straight across the horizon, and you can see the blue sky. Now image, and image as clearly as you can, a

huge, gigantic, circle in the sky in which two words are inscribed; the two words are 'Healing Energy'. Picture in your mind these two words, 'Healing Energy' inside that huge circle, which is *The Source of Healing*. Your thought is focused on the two words 'Healing Energy' only and nothing else. If any other thought comes to your mind, ignore it and bring back your visual attention ('imaged thought') to the words **'Healing Energy'**. Focus on the two words *Healing Energy*, for a few minutes; remain absorbed in them as vividly as you can. While in this state that you are, do the following:

3) With every inhalation (breathing in) feel that you are drawing Healing Energy from the Source of Energy; and with every exhalation (breathing out) feel that the energy you have drawn from the 'Source' is being spread throughout your body; believe and rest assured that every part, organ and even every cell of your body is being nourished with this Healing Energy. With every breathing 'in' and 'out' feel more vividly the spread and the flow of the *energy* in your body. Remember all this goes on while you are intensely imaging the two words, ***Healing Energy,*** inside the big circle in the sky. Do this (the exercise of breathing in and breathing out) for about ten minutes; however don't worry about time.

Now, if you have, a mal-functioning organ, or if you have a tumour in your body, or if you suffer, perhaps every now and then, from a pain in your body or from headache, then spend the last couple of minutes on focusing on that part of your body, i.e. 'showering' it (through breathing in and out) with *Healing Energy* and feel this energy flowing in that particular part of your body; feel it clearly and enjoy it vividly.

Well, now you are back, sitting on that straight-back chair, where you were when you started. This means no more imagery---you don't have to visualise anything. For a few minutes feel and enjoy the calmness of your mind and the state of relaxation that your body is in. Then, after that, open your eyes, move your hands and legs gently and then get up. Move about gently (not quickly or hastily) for a few minutes. Remember, what you have done is self-healing, healing the 'whole person'. In the process, you have been engaged in *reversing the cycle* of the development or process of any mal-functioning part of your body or any negative conditions that existed in you. As has been said this has to be practised regularly (every day), and also it has to be done purposefully and with commitment.

N.B. Over a period of time you will become quite good at practising this Self-Healing technique, and most probably you will also notice a reasonable degree of *'sense of well-being'* in yourself. Then, gradually, you may feel 'self-therapeutically creative' and you may like to give some "changes" to this method of self-healing. If so, please, do not alter the 'Basic Format' of this Self-Healing Method (the main ingredients, key feature, are *'breathing discipline'* and *'imaged thoughts'*). I wish you **All The Blessings and The Best of Self-Healing.**

15

Aramaic: The Blessed Language
of a Great (but now forgotten) People

"Not even the thousand years of Greek rule under Seleucide, Romans, and Byzantium were able to annihilate Aramaic as a language and Assyrian cultural identity from the Near East. ....
....The essential thing is that the Assyrians still preserve their ethnic, cultural and linguistic identity in spite of their loss of political power and heavy persecutions they have experienced especially in the Christian era." ('Assyrians After Assyria',

Professor Simo Parpola, Dept of Assyriology,
Helsinki University.)

General Perspectives of Aramaic Language

Aramaic is the ' linguistic heritage ' of the Assyrians. They inherited this precious language from their 'cousins', the Arameans. Assyrians and Arameans were the offspring of two brothers Ashur and Aram respectively-------the sons of Shem (Sem), from which the term 'Semitic' is derived. (Shem had five sons. Genesis, 10:22)

The previous language of the Assyrians and the Babylonians was Assyro-Babylonian (Akkadian), notorious for being a difficult language but well known for being a 'cultured language'. It was the first Semitic language, and it was written in cuneiform and spoken for nearly 3000 years in Mesopotamia ---- from around 3500 to 600 BC; it became extinct around 200 BC.

It was about 800 BC that Aramaic language began its 'linguistic migration' from Aramea to Assyria. Indeed, Aramaic was the indigenous language of the Arameans. The Kingdom of Aramea, which consisted of something like forty tribes, was adjacent to Assyria (Assyrian Empire). The spreading of the Aramaic language in Assyria was taking place very rapidly, permeating into Assyrian social strata. This linguistic spreading was so successful that in twenty years time Aramaic was officially accepted as the second language of Assyria. Encyclopaedia Britannica says, " By the eight century BC it [Aramaic language] became accepted by the Assyrians as the second language." (This was in my view about 770BC.) Aramaic continued its spreading successfully, in fact so much so that before 750 BC it supplanted the Akkadian language, thus giving an end to the 'linguistic rivalry' that was going on between Akkadian and Aramaic. (Some scholars believe such 'linguistic supplantation' took place probably before 800 BC) So, Aramaic became the official, the first, language of the Assyrians and the Assyrian Empire. At this stage, and for a few centuries thereafter, Aramaic language was referred to as the Imperial Aramaic. Not immediately (as most scholars in this field think), but after five decades or so Aramaic began gradually spreading out over Babylonia, the 'sister' of Assyria. It did not supplant the Akkadian language in Babylon quickly; this took place most probably not before 650 BC (say, between 650 and 640 BC).

At this stage it would be sheer folly not to say also a few words about Sumerian language, known as the earliest language in Mesopotamia (most philologists say, 'in the world'). This language, Sumerian, is said to have been rich in epics, hymns, rituals etc. Here too, more or less, is the same story. Akkadian gradually became the second language of Sumerian people and then, over a period of time, it supplanted their language. Sumeria dates back 7000 years. This ancient People, without any doubt, left some of her cultural influence in Mesopotamia.

From 600 BC Aramaic began to spread out beyond the boundaries of Mesopotamia. The whole process of this ' linguistic migration' was so fast that in less than a century Aramaic became the 'lingua franca' of most (nearly all) Middle Eastern countries. This 'first lingua franca in the world' was also welcomed in some other parts of the world. The most welcoming nation to this "amazing language" was probably the Persian Nation and its great Empire. Encyclopaedia Britannica says, "...It [Aramaic} subsequently became the official language of Achaemian Persian dynasty (550 - 330)'. Aramaic enjoyed its status as lingua franca for more than a thousand years. The "ubiquitous status" of this extraordinary language began diminishing in the late 8th century A.D, and such diminution was in direct proportion to the speed of its being replaced with Arabic language.

I have referred to Aramaic as 'amazing language', and in doing so I have not indulged in exaggeration; I simply meant what I said. This precious language, Leeshana Aramayah, includes elegant idiomatic expressions, rich metaphors and similes, meaningful paradoxes, amusing hyperbole, profound parables, and so on----
--but! But there is something more than these that may render

Aramaic unique among all the languages in the world. And that is the 'higher dimensional aspect' of it------something of a Spiritual Dimension. Aramaic is a 'mystical' language, and in certain aspects it is completely different from other languages. It is a language of the Soul/Mind..

Robert Allen, The President of The Aramaic Bible Society says,

"Aramaic is a language which inherently expresses the manner in which the mind functions."

Absolutely true! It could be said that it is the language of 'Here and Now'. In Aramaic the 'divisions' between Past, Present, and Future are not 'thick concrete walls' as they are in other languages. (Here I am not referring to grammatical tenses as such but to the *Spiritual Aspect of Time* ---- Here and Now.) Thus, Subject and Object in Aramaic seem to be on the same 'spectrum', and Objective and Subjective to be not as 'opposites' but on the same 'continuum'. It is the Language of Eternity-----"I Am". Is it any wonder then why Western Aramaic scholars have never been able to grasp (fully understand) this "Spiritual Beauty", the 'non-physical' aspect, of Aramaic language? Those who were born in and with this language can appreciate and discern such Beauty far better than those who learn it. Now we can see why the late Doctor George Lamsa (1892-1975) an Assyrian, had by far a deeper knowledge of Aramaic and particularly a more profound insight into Aramaic Bible (and biblical Scriptures as a whole) than any Western Aramaic scholar. These scholars, in general, did not like him and many of them showed unkind attitude to him. This was mainly, but not only, because this Assyrian Scholar of Bible and Aramaic language, Doctor George Lamsa, often in his writings would refer to Aramaic as "The language of the Assyrians" and also sometimes as "The language of The Church of The East". Most, but not all, Western Aramaic scholars did not like this kind of statements. As far as the West is concerned, the Assyrians do not exist anymore. However, after the fall of their Empire in 612 BC, a large number of the Assyrians survived, and went on building a new world of their own. To cut the story short, Assyrians after Assyria have contributed to the civilization of the world not less but 'possibly more' than when they had their glorious empire, Moreover, when it comes to the spreading of Christianity in the World, the Assyrian Church of The East, through her dedicated Assyrian Missionaries, is second to none. Competent and honest Assyriologists would attest the things that I have stated here. Doctor Lamsa, who had a profound knowledge of the history of his people, wrote on and spoke of the immense contributions of the Assyrians to the World.

Furthermore, Lamsa always maintained, and never doubted, that bible was originally written in the language spoken by Christ Jesus, i.e. Aramaic. On the translations of the Old and New Testaments, this spiritually adventurous man, Doctor Lamsa, spent more than twenty years. In his books, articles, and lectures, Lamsa always maintained that Aramaic Bibles are superior sources of Biblical truth compared with Latin or Greek based Texts. His writings are illuminating and his explanations are elucidating. He was a true Scholar and a truly learned man. He spoke at least seven languages. .

I wonder if the Western scholars of Aramaic realize how great has been the contribution made by the Assyrians and their scholars (in terms of both amount and quality) to the Aramaic language and literature. These contributors also include some of the great theologians of the Assyrian Church of The East whose ' Missionary Enterprise' still remains unparalleled in spreading Christianity in the world.

Assyrians love their language; they think of it as their ' highly important cultural identity', and also as the "Connecting Link" between them, as a People, and their Saviour who spoke (did His

Teachings and Preaching) in this language, Aramaic. And that is why, as the heirs to this precious language, the Assyrians have taken 'Good Care' of their language, and their Scholars have developed it linguistically and enriched it philologically. Such enrichment also includes the suffusion of many Akkadian words in Aramaic language and literature. And let me also say that more than two-third (at least 70 percent) of the treasure of Aramaic literature available today is the product of the literary works of the Assyrian writers, including many of their eminent scholars.

Later, the city that became the place for the development of Aramaic was Edessa (in Assyrian/Aramaic language called 'Urhai', today it is part of Turkey referred to as 'Urfa'). Edessa, especially in the 5th century A.D, was renowned for being the cultural and intellectual centre of the Assyrians. The Church of Rome could not stand this. Thus, Assyrian scholars and intellectuals and their Church (The Church of The East) were ordered to get out of Edessa, which was at the time part of the Roman Empire. Assyrians had to close down their colleges and educational centers and leave. So, they moved to Nisibin, which was another cultural city of the Assyrians and not very far from Edessa. This city was under the authority of the Persian Empire. It was here, in Nisibin, modelled on Edessa, that the Assyrians established *The First University In The World*. The University consisted of several Faculties: Philosophy and Logic, Theology, Literature and Philology, Science, Mathematics, and also Medicine.. At this stage (period in history) Aramaic language and literature were flourishing, at their peak...

Here, I should like to recommend to the readers of this essay, particularly to those from the West, to read the book, by the late Benjamine Wilkinson, titled, *Truth Triumphant: The Church in The Wilderness*. In this book we read about the great contributions of the Assyrians after Assyria to so many parts of the world. These contributions were made mostly under the authority/ Leadership of their Church, The Church of The East. The contributions of this Holy Institution, the Assyrian Church of The East, have truly been incredible. In the beginning of chapter 23 Wilkinson says,

> **"Japan owes much of her civilization to The Church of The East. This may come as a surprise to many. If so, there will be more surprises in the store for those who are not informed as to how strong a determining factor Christianity was in the corner of Island Empire."**

Also, the author somewhere in chapter 24 says,

> **"It has been noted how in 9th century the civilizing education system of the Church of The East determined the Golden Age of the mighty Arabian Empire so much as that it permeated the literature of China and Japan in the East and paved the way for the founding of universities in Europe."**

Furthermore, Assyrians also through their Aramaic language and literature have contributed immensely to the language and education of many Peoples. Paul Younan, an Aramaic scholar and the translator (interlinear translation) of the New Testament from Peshitta Aramaic into English, in his interesting and informative Paper, 'On The Preservation and the Advancement of Aramaic Language in the age of Internet', says,

> **" Aramaic language encompasses a treasure of nearly 3000 years of Assyrian civilization including our versions of the Bible (The Peshitta Aramaic), hymns and poems, translations of Greek works, biblical commentaries, historical works, compilations of lives of Saints, and works on Philosophy, Grammar, Medicine and Science."**

Who knows? Perhaps the Assyrians ' were meant' to prepare the Aramaic language and to *Pave The Way* (philologically) for The Coming of The Lord and Saviour, Who Spoke This Spiritual Language. Who knows!

Indeed, the treasure of Aramaic language is priceless. Assyrian Aramaic became for centuries the 'source of enrichment' for the development of many languages in particular Arabic and Hebrew. In fact Hebrew language emerged as a dialect from Aramaic language and then gradually, as many dialects do, became a language in its own right. Arabic language, too, evolved from Aramaic language. (Incidentally, Arabic is not an ancient language. Its earliest Texts go back not even 1500 years) And what about Greek? Though not from the same 'Linguistic Family' (Semitic), Greek, philologically, stemmed in part from Aramaic language; this enriched immensely Greek language philosophically, theologically, and in terms of literature. What about the Aramaic Alphabet? In the same Paper, Paul Younan says,

> **"Aramaic was the first lingua franca. It also served as a basis for almost every Western-based alphabet in existence, from Greek to English and from Hebrew to Arabic as well as various Asian scripts used in India and Mongolia."**

Today there are nearly (if not quite) twenty variants of Aramaic language, which are referred to by some students of Aramaic as 'languages'. This makes nonsense of what is meant by "language'; they are not languages. Nor are they all (as most scholars say) 'dialects'. Some of them are dialects, and some of them are 'versions' (Forms) of Aramaic. They all of course come under the 'umbrella' of *Aramaic language*. Until about 250 BC there was probably one Aramaic, the Assyrian (Imperial) Aramaic. At the time of Jesus apart from this original Aramaic there were two or three variants of this language. The real diversification of Aramaic took place from around 1000 to 1500 AD. Only one or perhaps two of the variants of Aramaic available today are Western Aramaic, the rest are all Eastern ---- Eastern Aramaic. (Here, the terms ' Eastern' and ' Western' refer to certain geographical regions inhabited by Aramaic speakers in Middle East at that time.)

At this point I find it fitting as well as necessary to talk briefly about the 'Library of Nineveh', which was the creation of one of the Greatest Kings of the Assyrian empire, Ashur Banipal, who ruled Assyria from 669 to 627 BC. That is why it is also referred to as the 'Library of Ashur Banipal'. It was discovered by an English archaeological adventurer called Henry Layard in 1849. These lovers of words, language and literature, the Assyrians, in particular their scholars and intellectuals under the creative leadership of the King set up an extraordinary library in their Capital City of Nineveh. It contained more than 30,000 original documents, all cuneiform tablets containing more than one thousand distinct texts of highest standard and significance. .

I am going to quote from a paper, *Ashur Banipal: The Warrior-Librarian*, which may at least to some degree explain the Library's standard and significance that I just alluded to. It says,

> **"With the creation of the library at Nineveh, Ashur Banipal ensured not only the preservation of his culture and society but laid the foundations for a form of preservation used by following cultures. The content of his library provided the basis for many national libraries that followed. It became important to not only preserve archives and records, but also national stories and mythological works as well. The importance of Ashur Banipal's library cannot be overstated. The collection was spread out into many rooms according to subject matter. Some rooms were devoted to history and government, others to religion and magic and still others to geography, science, poetry, etc. The library even stored what could be called government classified material. ....."....**

After the death of Ashur Banipal in 627 BC Assyria began gradually declining. It was, however, in 612 BC that the beautiful and majestic city of Nineveh was destroyed, and many parts of it, including the Library of Ashur Banipal, were burnt down by the enemies of the Assyria, chiefly by Medes and Babylonia

The Bible and The Aramaic Language

"Many of our Assyrian people in America unfortunately are unaware of the key which we hold to the Scriptures and especially to the New Testament through the Aramaic language that our Lord Jesus Christ spoke and preached during His Mission on Earth."

(From The Patriarch of
The Church of The East,
His Holiness the late
Mar Shimun------in an
address to the Assyrians in
Modesto, California, USA)

The Holy Text, the official text, of the Assyrian Church of The East is the *Pshitta Aramaic*, which is the rich and authoritative Version of Aramaic language. 'Pshitta' means 'simple', 'straight', 'uncomplicated', 'sincere'; it would be reasonable to infer from the combination of these meanings that here the word 'Pshitta' denotes 'genuineness'-------genuine Form, Version, of Aramaic. (Of course this does not mean that other variants of the language in question are not genuine.)

As regards to the holding of "The Key to the Scriptures", credit ought to be given to The Assyrian Church of The East and the martyrs thereof who throughout gave their lives for the preservation of the original scriptures handed down by the Apostles, from one generation to another, kept in the security of The Church. Let me again quote from His Holiness, the late Mar Shimun from His letter to a publisher on 4th April 1957-------He says,

".................as the Patriarch and the Head of

the Holy Apostolic and Catholic Church of The
East, we wish to state that the Church of The
East received the Scriptures from the hands of
the blessed Apostles themselves in the original
language spoken by Our Lord Jesus Christ Himself and that Peshitta [Pshitta Aramaic] is the
Text of that Church which has come down from
the biblical times without any change or revision."

Indeed, Peshitta Aramaic, as the Version of the Holy Text of
the Church, is the highly developed Form (Version) of the original
Aramaic, and also it is the most authoritative. And contrary to what
many Western biblical scholars believe, the Four Gospels were
written, completed, only some years after the Ascension of Lord
Jesus to the Mystical Eternity of The Heavenly Father. Lamsa says,

" ....that is also true of the Gospels.
They [The Gospels] were written a few
years after the Resurrection, and some
of the portions were written by Matthew
while Jesus preaching. They were not
handed down orally and then written after
Pauline Epistles as some Western
scholars say; they were written years
before those Epistles."

Assyrians, as a People, were the first to believe in Jesus as the Lord
and Saviour. After the Assyrians, in my view, it was the Chaldeans
who followed Jesus. However, Victor Alexander, an Aramaic Scholar,
slightly differs from this view. He says,

"The Ashurai Nation [Assyrian Nation]
became Christian in the first century,
followed by the Armenians and Chaldeans."

Assyrians, while The Lord was doing His Teachings and
Preachings on the Earth, were preparing for the establishment of
the First Apostolic Church. Such 'Holy Establishment' took place

in Seluquia-Ctesiphon, Babylon, in 33 AD. The Founders of this 'First Apostolic Church' were Thomas and Bartholomew (from the Twelve Disciples), and Mar Addai and Mar Mari (from the Seventy Followers). This Church in Babylon, (which later was named The Holy Apostolic Church of the East), is mentioned by Shimun Keepa (Simon Peter) in one of his Epistles. He Says,

"The chosen Church that is in Babylon,
elected together with you, saluteth you.
And so does my son Marcus."
(1Peter, 5:13)

This Church, The Church of the East, today is usually referred to as the Assyrian Church of The East. The present Patriarch and The Head of The Church is His Holiness Mar Dinkha. The rituals and the liturgy of the Church are carried out in the spiritually rich and beautiful version of Aramaic that Our Saviour spoke and also used in His Teaching and Preaching, The Master was the master of both the use and the usage of the words in Aramaic Language. It is only in Aramaic language that The Lord's Prayer rhymes, and it is only in Aramaic Bible (only in Aramaic language) that The Beatitudes (Sermon on The Mount) rhyme. In fact there are hundreds of poetical verses in Aramaic Bible. Such rhyming, rhythmic beauty and poetical skill and word-play in Aramaic New Testament cannot be translated into any other language. Now we can see why this 'philological beauty' is absent in Greek bible, which is a 'not-quite-good' translation of the Aramaic Bible.

There is no shred of evidence that the Bible was originally written in Greek; it has only been taken for granted ----- and that is all. Today the Aramaic Primacy is growing more and more ---- -------- the view that the Bible was originally written in Aramaic (Pshitta Aramaic). Yes indeed, New Testament was first penned in the language of Eshoo Msheekha (Jesus Christ). That the Bible was first written in Greek is not just a misconception but also a wishful thinking. The Greeks propagated this 'Myth' and the West gladly followed this 'mythical concept'. What is more 'amusingly interesting' is the fact that there are already a few web-pages on the internet which say that Jesus also knew Greek, and that his disciples too were quite familiar with Greek language; thus Jesus and His disciples, they say, sometimes used to communicate in Greek! This is Absolutely Nonsense! **Doctor George Lamsa, says,**

> **" Not even a word of Scriptures was written in Greek...**
> **............Jesus and his disciples not only could not speak Greek but they never heard it spoken."**

And let me say here (since I passingly made a reference to civilization) that without the civilization of Mesopotamia the Greek and Persian civilizations couldn't reach the standard that they reached. Both Persians and Greeks owe a great deal of their civilizations to the Assyrians. Every Assyriologist and every competent and honest Scholar of the Ancient History will attest this. And what about the Romans? The eminent Assyriologist of Helsinki University, Simo Parpola, in his well-known Paper, 'Assyrians After Assyria', says,

> **"We know that Greeks and Romans from Plato till antiquities kept learning spirituality and science from the Assyrians and the Babylonians."**

Yes indeed, also the Romans and the Roman Empire learned, and learned a lot, from the Assyrians and Babylonians, the inhabitants of Mesopotamia (The Cradle of Civilization).

Jesus, Aramaic Language, and The Assyrians

Now, let us see: what kind or form (Version) of Aramaic did Lord Jesus Speak? On this question most biblical scholars say that it was "Galilean". However, they each say it in their own way. These are the most stated sentences on this issue. They are,

> **"Jesus spoke Galilean dialect of Aramaic."**
> **"Jesus spoke Aramaic with Galilean accent."**
> **"Jesus spoke Galilean Aramaic."**

The three above statements denote the same thing: Jesus spoke the Aramaic that was spoken in Galilee, which was a region, province, in Palestine. So, let us know more about this 'Special Region' of Palestine. And who can describe it better than the man himself, Doctor G. Lamsa. In his book 'The Kingdom on Earth ', he says,

> **" Indeed, Galilee is a beautiful and luxuriant**
> **land when compared with the arid hills of**
> **Judea and adjacent lands of.... Nature has**
> **graced this land with many scenic beauties**
> **and abundant water in a part of the world that**
> **water is scarce and precious. There is some**
> **thing mysterious about nature around the**
> **lake of Galilee.......Galilee was far away from**
> **Jerusalem, the centre of Jewish religion....The**
> **Galileans were free-thinkers, free from the**
> **complex system of .... Therefore, they [the**
> **Galileans] were eager for the fulfilment of the**
> **Messianic promises, and they prayed with**
> **vehement supplications for the deliverance of**
> **their lands from the pagan rulers."**

In Palestine, at the time, there were only ' a couple of ' variants of Aramaic, and the best (the richest) one was spoken by Galilean people. Nazareth was a town in Galilee. Jesus was from this town----------hence, Jesus of Nazareth. Galileans, originally, were Ashurayeh (Assyrians) who centuries before had moved, immigrated, to that region of Palestine. Victor Alexander, who has translated the New Testament from Aramaic into English (and he is busy translating the Old Testament too), says that not only were the Galileans Ashurayeh, but also their language, Galilean Aramaic, is the key to the Scriptures, or as he puts it ' to the translation of the Bible'. Let us see what he says.

> **"......the language that Jesus spoke with**
> **the Galilean accent...........The Galileans**
> **were originally from Nineveh........Nineveh**
> **was the Capital of Ashur[Assyria]. Ashurai**
> **[Assyrian] language in the Galilean dialect**
> **is the key to the translation of the Bible."**

At the same time we have many biblical scholars particularly 'Aramaic Primacists' who maintain that the version called ' Peshitta Aramaic' (which is the official, the Holy, Text of the Apostolic Church of The East), was the language of Jesus and His Disciples, and it is ' The Key to the Scriptures'. In fact this was emphasized (more than once) by His Holiness the late Mar Shimoon, the Patriarch of the Assyrian Church of The East (the one before the present Patriarch), who was also an eminent Scholar of the Bible and the Aramaic language; I have already quoted from Him.

From what has been said I draw my own conclusion, which in part is implicit in what has been said. As I see it both Peshitta Aramaic and Galilean Aramaic were Aramayah d'Ashurayeh, Assyrian Aramaic, but with a difference. The difference is this: Peshitta Aramaic is the Special Version (the Literary Form) and the Galilean Aramaic refers to the Common Speech (the vernacular) of Aramayah d'Ashurayeh. Put another way, Galilean was the 'Colloquialism' (Dialect) of Peshitta. (I prefer to write the name of the version in question as 'Pshitta' since in Aramaic it is a two-syllabic word, but many writers including some scholars write it as 'Peshitta')

Considering what has been said in this section, we can see the close affinity between the Ashurayeh (Assyrians) and our Lord Jesus. He, The Lord, grew up and lived among them and spoke their language. Some of the most faithful friends and closest followers of Msheekha (Christ) were Assyrians, or as He called them, 'The men of Nineveh'. He loved them, that is why He gave them *Hope and Encouragement* when He said,

"The men of Nineveh [The Assyrians] shall rise
in judgement with this generation and shall condemn
it because they repented at the preaching
of Jona; and behold a greater than Jona is here."
(Matthew, 12: :41)

Isn't it the 'Greatest Blessing' for a people or a nation to be pronounced 'repented' by Jesus Christ?

Assyrians usually refer to their language as Leeshana Ashuraya (Assyrian Language). And why not? They have spoken this language for nearly 3000 years; therefore they have every right to think of it and to call it 'Assyrian language'. There have been, anyhow, people including some philological students who have used the two terms ' Ashurai ' and ' Aramaic ' interchangeably. Perhaps this has been more prevalent among the Jews particularly in relation to the Hebrew Scriptures. Victor Alexander says,

"There is another name for ancient Aramaic.
The Jewish scholars of scriptures today talk
of the Ashuri language and they call the sacred
language of the Torah "Ashurit".........Ktave
Ashuri or Ashuri writing. This is the language
in which the Ten Commandments were written,
and the only sacred language of the Old Testament
according to most Jewish scholars."

Such usage of the two words, 'Ashurai and Aramaic' (using them interchangeably), is a reminder of the ancient Assyrians and the ancient Aramaic, and also of the Ashurayeh and their language at the time of our Lord in Palestine.

Reverend Doctor R.A.Wigram, an eminent English Scholar of theology and ecclesiastical history (who in the early part of the 20th century wrote several books on the Assyrians and on The Church of The East, and related areas) in his book, ' The cradle of Mankind' says,

"One thing is certain that the Assyrians boast with justice that they alone of all Christian Nations still keep as their language which is acknowledged to be the language of Palestine [referring to Galilee]."

Apart from the Assyrians there are in the Middle East the Chaldeans, and also one or two other communities who speak Aramaic language, The versions or dialects of Aramaic spoken today are usually referred to, especially by the linguists and philologists in the West, as Modern Aramaic or Neo-Aramaic. Languages, of course, in the course of time, change as a 'linguistic evolutionary process'. However, considering the vicissitudes over the past centuries that the Aramaic speaking people, particularly the Assyrians, have been through, we can boldly say that it is a 'not-much-changed' language---------Still Spiritually Rich Aramaic!

Incidentally, the Chaldeans of today most (if not all) of them are the Assyrians who became the followers of the Roman Catholic Church. Let me tell a long story in a few lines only.

In 1550s AD a number of Assyrians headed by a monk by the name of Sulaqa broke away from The Church of The East. Sulaqa together with his 'entourage' went to Rome and visited the Pope. He was warmly welcomed and was gladly appointed the 'Patriarch of the new Uniate' by the then Pope, Julias the third, who also designated the new Church 'the Chaldean Church' in order to have it distinguished, separated, from the Assyrian Church of The East.

Thus, all the Assyrian followers of Sulaqa betrayed their ecclesiastical identity and denied their national entity; they followed the Church of Rome and called themselves "Chaldeans'. And even to this day if or when an Assyrian joins Roman Catholic Church he or she (most probably) will call himself or herself "a Chaldean", which is absolutely absurd, and also sad --- sad indeed! .

Today there are over three millions (nearly 3.5 M) Assyrians scattered almost in every corner of the world. Almost two-third of them are in Iraq, America and Syria, and the rest of them, mostly (but not all), are in Europe. These inherently cultured and instinctively courageous people (despite having been through the worst kinds of persecutions throughout the centuries, and in spite of having been Stateless since 612BC, and also having experienced the horror, grief and sorrows of the killings, massacres, of their Assyrian brothers and sisters, most of which took place in the 20th century), have preserved their language, and have faithfully kept their Christianity and their Ancient Church, and they have also upheld their traditions and cultural values. And above all, these great people, the Assyrians, have never stopped contributing to the well being of their fellow human beings and to the World that they live in

Reference:

The amount of literature on Aramaic Language on the Internet is huge. And also there is plenty of material on obtaining knowledge and information on this language in relation to the Assyrians and The Church of The East, and on the interesting topic of 'Aramaic Primacy'. However, I am going to supply the reader with a few references that are directly related to some of the Quotations and/or closely linked to a few areas dealt with in this Article.

Paul Younan's Website: **www.peshitta.org**

(This website is unique in its kind. It contains so many things and it deals with so many issues: Aramaic, Bible, The Church of

The East, Forum Debates, Teaching Assyrian Aramaic, and Books and Articles on such areas. Paul's Interlinear Translation of the New Testament, from peshitta Aramaic into English is available on this Website. And his Paper, 'Preservation and Advancement of the Aramaic Language', from which I have quoted, is accessible on Paul Younan's Website.)

Victor Alexander's Website: **www.v-a.com/bible/aramaic.html**

www.v-a.com/bible/aramaic_project_review_index.html

(Victor's Articles are all interesting, informative and here and there with 'some originality'. He writes courageously. Alexander's translated Bible from Aramaic into English also is available on his Website.)

Doctor George Lamsa. The Websites that deal with Lamsa and his Works (Books etc) are mostly part of The Aramaic Bible Society.

www.metamind.net

www.aramaic.org

www.aramaicbiblecenter.com

www.aramaic.org/AWAKE.html

www.metamind.net/Aboarder.html

www.aramaic.org/News.html

And here are three books by George Lamsa:

Holy Bible (Peshitta), Harper, San Francisco 1933

New Testament Origin, Publisher 'Aramaic Bible Society' Sep. 2003

The Oldest Christian People, Georgis Press, Dec. 2006

Professor Simo Parpola's Paper, 'Assyrians After Assyria'

www.atour.education/20000703a.html

There is a Compiled Work on PESHITTA ARAMAIC PRIMACY ----compiled and discussed by Chris Lancaster. I highly recommend it. The website address is, **www.peshitta.netfirms.com** (Then, Click on "Was The New Testament Really Written In Greek?') And here is another website administered by Chris: **www.truth777.netfirms.com**

Sam Razali's Website: on 'Assyrians' **www.assyrians-homeland.org**

also **www.humandynamicwholeness.org** (primarily his professional website)

Benjamin Wilkinson, 'Truth Triumphant: :The Church in The Wilderness', Pub. Teach Services Inc. U.S.A

Doctor W.A.Wigram wrote several books on Assyrians and The Church of The East, and related areas. One of his books is The History of The Church of The East. Furthermore, there is, in this article, a Quotation from one of Wigram's books, 'The Cradle of The Mankind' which the reader may find and read on Paul Younan's Website.

Finally, two more books ------- the second one (by Rev. Warda) is also available on P.Younan's website.

Saggs, P.Y. The Might that was Assyria

Warda, Rev. Joel, E. The Flickering Light of Asia

16

Science and Spirit

Let me first of all say what the word 'Spirit' means or refers to in the phrase 'Science and Spirit' which is the title of this essay. The term 'Spirit' in this context refers to that 'realm' which transcends 'matter' and all materially related things and is beyond this phenomenal world. Also, the word 'spirit' here may refer to 'the enthusiasm and great interest' of a person in knowing about the 'realm' in question. The word 'Spirit' may also (some times and in certain contexts in this discussion) denote the 'followers of the Spirit' as 'Science' might stand for 'Scientists' (or, the 'followers of the Science').

The bringing of science and spirit together and the endeavour to narrow the gap between them has for the past thirty years or so been of a great interest among an increasing number of scientists from various fields of knowledge. This has been happening throughout the world, but far more so in the West for the obvious reasons. ('Obvious' because the West is by far more materialistic and non-spiritual than the East.)

During this period things have been moving on and some changes have taken place. In short, gradual achievements are in progress. The gap is still wide ---- as wide as a valley. We still have on one side *Newtonian mechanistic Model of Reality* with all its related concepts and theories, and on the other side of the 'valley' is and has always been the 'Spirit' and its related philosophies, metaphysics, religions, theologies, and so on..

As has been said, across the 'valley' (the huge gap between Science and Spirit) things have been happening. More and more scientists (from all fields of science), philosophers, psychologists, physicists and other learned people are becoming disillusioned ------- disillusioned on good grounds. They see the mechanistic concept of the universe as totally bankrupt; they see it as a concept consisting mainly of purposeless mechanical paradigms, which are stuck and cannot go further because they are both limited and limiting.

These disillusioned scientists have welcomed Einstein's *Model of Reality* central to which is his concept of matter and energy based on his well-known equation, $E=mc^2$ (the most famous scientific equation of the 20th century). This non-mechanistic and vibrant paradigm tells us that matter is another form of energy; moreover, mass and energy are convertible. This is incredible: compare this with the solid, lifeless, and mechanistic (classical) model of reality.

Then along came Quantum physics, better known as Quantum Mechanics, central to which is Wave-Particle duality. This popular 'duality' debunked the belief that matter consists of certain 'solid building blocks'. Indeed, it went much further than Einsteinian model. Einstein's physics shook the foundation of the mechanistic structure of Newton's theory of the universe. Quantum physics brought down completely the fortifying fences of the classical physics and demolished the whole structure of Newtonian model of reality which had been in existence for around 300 years. *These revolutionary theories and vibrant paradigms, the New Physics as a whole, tore down the 'three-centuries-veil' that had practically blinded most scientists and many philosophers.* Now they are beginning to see the 'reality', the scientific truth, which, let us hope, will be followed by seeing 'The Truth'.

Today there are more than 200 societies, associations, organizations and networks in the world which all aim at understanding the differences (attitudinal, philosophical, conceptual, spiritual etc) that has kept science and religion apart for centuries. They are trying to lessen, as far as feasible, the 'estrangement' that has existed for so long between them. These societies and networks have members from a wide range of fields of science, and areas of knowledge. I am glad to say that I am a member of a couple of them. We, the members, are doing our best to see Science and Spirit getting closer and closer. The more we succeed in achieving such aim the greater will be the possibility of the 'integration' of the two, which is our ultimate goal. Am I getting carried away, or indulging in 'euphoria'?.

Whether indulging in 'euphoria' or 'utopia', I am sure it is feasible. After all, both groups, the scientists and the followers of the Spirit, are the seekers of the 'truth', but with a difference. The difference between them is that the former tries to find 'the truth' through the process of scientific researches, methodologies, experiments and analyses in the 'physical world'; the latter is in search of 'the truth' through direct experiences of introspective and personal nature in the non-physical world. Put another way, one group tries to understand things and to find the truth about their physical nature, mainly by disentangling the physical complexities through scientific means The other group, however, tries to know the 'truth' about the things of non-physical nature, by carrying out certain disciplines and spiritual practices ---- uncovering the metaphysical complexities through introspection, personal experiences and spiritual insights. Let us hope such practices (though applied in different ways) *for finding the truth* will progressively enable the two groups of 'Science' and 'Spirit' to approach one another more frequently and especially more amicably. In fact at present this is gradually happening. The contingency of such 'integration', though not feasible in near future, is not remote.

Central to the advancement of the field of Science and Spirit is the 'spirit' (high enthusiasm and great interest) of the students and researchers in the field in question. Such spirit is dynamic and it arises from the 'center of being' of the person. It is an 'awareness' related to *"Human Dynamic Wholeness"* (which is the title of my *Cosmic Concept of Human being,* and also it is the name of my Website dedicated to *'Science and Spirit'.* . This, what I have called 'awareness' (or 'realization') stems from the 'Center of Being'. In other words, it is one's Awareness of one's Self, or Soul. Indeed, I have always maintained that the 'chief element' in narrowing the gap between Science and Spirit depends mainly on the *Awareness of the Spirit Within* among the scientists. Expressed differently, it is a 'Transformation' taking place in the Consciousness of people in general and the scientists in particular. When this happens the 'Scientific arrogance' will melt away and the 'spiritual pride' (and I hope also religious bigotry) will die away. Groundless doubts will diminish, and the silly cynicisms will rapidly disappear.

The notion that science and spirit cannot mix is false, and the idea that the two cannot exist side by side is totally wrong. Science and Spirit can, and will, get along with each other amicably and can 'cohabitate' in a most civilized manner. In truth, they need each other, though in different ways. **Science will glorify the Spirit, Spirit will magnify the Science.** *Spirit will widen the horizons of Science, and will awaken the scientists to the*

Realities of Knowledge that they could not 'see' before. Both Scientific and Spiritual fields of knowledge, originate from the same Source and belong to the same Truth. Spirit is The Giver of Life to Matter, and Matter gives Form to Spirit. Spirit is manifested through Matter, Matter 'comes to being' through Spirit.

Supposing the 'full integration' of Science and Spirit has taken place. What will be the result of this long-awaited integration? What is going to be the consequence of such 'matrimony' (the 'marriage' of Science and Spirit) as far as the fields of knowledge are concerned? What changes (if any) will come about; and if so, in what areas of knowledge such changes or improvements will take place? Let me try to deal with these questions in one or two short paragraphs.

Quantum mechanics will not stay as it is. Another 'New Physics' is desperately needed. This New Physics will be a combination of 'restructured' Quantum theory and 'reformulated' Einstein's physics chiefly his theories of relativity. Such *restructuring and reformulation* in the science of physics are absolutely essential in order to have a *'Cohesive Physics'.* . And let's call it 'Cohesive Physics'; it is an appropriate title for the physics of the 21st century. Let us hope it will catch on and will be accepted as the proper title for the future new physics.

In the 21st century Consciousness will be on top of the agenda for investigation among physicists and psychologists. Another area of investigation will be the 'Subtle Energies', both those of human being's (human bioenergies) and of the universe (cosmic energy). In fact as I 'see' it, most part of this century, as far as Science is concerned, will be mainly research into the Subtle Energies, Consciousness and the study of the Universe. However, I see another area (in fact a 'controversial one') that will draw the attention of the scientists to itself; and that is *Mind-Brain Problem.* The Scientists, from different fields of knowledge, will come to realization that Mind and Body (mind and brain) work interdependently and as 'one whole', yet, it is the Mind that has the 'Primacy' over the Brain (read my theory of Mind-Brain Problem). Indeed, time will come (in fact it is coming) that even the most atheistically-minded biologists and the staunchest physicalist/materialist scientist will accept that **in Human Being, between Mind and Body, Mind is The Primary, even Supreme.**.

When no longer we see a chasm between Science and Spirit we shall have a far better understanding of life in general, and much healthier approach to our living in this world in particular, furthermore, we shall have. a clearer perception of our existence in this universe. The basis for all these is that we come to realization and face the acceptance of "the truth" that both Science and Spirit are of one *Source*. In order for all these to happen 'Awakening of the Consciousness to the Spirit' is needed; such need is more desperately felt (more urgent) among the scientists. This is already happening. Here I am reminded of that controversial German Catholic theologian, Hans kung. In one of his books, which he wrote many years ago, Hans kung says that the day will come when the scientist will say, *"Of course I believe in Spirit; I am a scientist"*. ***How right he was!***